Macromolecular Solutions

Pergamon Titles of Related Interest

Coetzee RECOMMENDED METHODS FOR PURIFICATION OF
SOLVENTS
Loening LIST OF STANDARD ABBREVIATIONS (SYMBOLS) FOR
SYNTHETIC POLYMERS AND POLYMER MATERIALS - BASIC
DEFINITIONS OF TERMS RELATING TO POLYMERS
Marcus, et al. COMPOUND FORMING EXTRACTANTS, SOLVATING
SOLVENTS AND INERT SOLVENTS, Vol. 15 of Chemical
Data Series
Muzzarelli NATURAL CHELATING POLYMERS
Nakajima MACROMOLECULAR CHEMISTRY, Vol. 5 of 26th Interna-
tional Congress of Pure and Applied Chemistry, Tokyo, 1977
Sedlacek MACROMOLECULAR MICROSYMPOSIA 16,
16th Microsymposium on Macromolecules, Prague, 1976

Related Journals*

CHROMOTOGRAPHIA
EUROPEAN POLYMER JOURNAL
ORGANIC GEOCHEMISTRY
PROGRESS IN POLYMER SCIENCE
SOLUBILITY DATA SERIES

*Free specimen copies available upon request.

Macromolecular Solutions
Solvent-Property
Relationships in Polymers

Edited by
Raymond B. Seymour
University of Southern Mississippi
G. Allan Stahl
Phillips Petroleum Company

PERGAMON PRESS
New York   Oxford   Toronto   Sydney   Paris   Frankfurt

6654-0756

CHEMISTRY

Pergamon Press Offices:

U.S.A.   Pergamon Press Inc., Maxwell House, Fairview Park,
                         Elmsford, New York 10523, U.S.A.

U.K.   Pergamon Press Ltd., Headington Hill Hall,
                         Oxford OX3 0BW, England

CANADA   Pergamon Press Canada Ltd., Suite 104, 150 Consumers Road,
                         Willowdale, Ontario M2J 1P9, Canada

AUSTRALIA   Pergamon Press (Aust.) Pty. Ltd., P.O. Box 544,
                         Potts Point, NSW 2011, Australia

FRANCE   Pergamon Press SARL, 24 rue des Ecoles,
                         75240 Paris, Cedex 05, France

FEDERAL REPUBLIC   Pergamon Press GmbH, Hammerweg 6
OF GERMANY   6242 Kronberg/Taunus, Federal Republic of Germany

Copyright © 1982 Pergamon Press Inc.

Library of Congress Cataloging in Publication Data

Main entry under title:

Macromolecular solutions.

    Includes papers presented at a symposium held by the American Chemical
Society at New York City, Aug. 23-28, 1981.
    1. Macromolecules--Congresses. 2. Solution (Chemistry)--Congresses. I.
Seymour, Raymond Benedict, 1912-
II. Stahl, G. Allan. III. American Chemical Society.
QD380.M28   1982   547.7'0454   81-23535
ISBN 0-8-026337-2   AACR2

Printed in the United States of America

DEDICATION

Our previous book on "Structure-Solubility Relationships in Polymers" was dedicated to Dr. Joel H. Hildebrand who was a pioneer in this field and the author of the first chapter in that book. Another pioneer in this field, Dr. Maurice Huggins, has written a chapter for this book.

In the absence of theoretical studies by pioneers such as Drs. Hildebrand and Huggins, there would be little theory on plastics versus solvents and probably not enough information to justify books on the subject.

It is our pleasure to dedicate this new book to Dr. Maurice L. Huggins.

Raymond B. Seymour

G. Allan Stahl

v

TABLE OF CONTENTS

The papers contained in this book were presented at a symposium on "Macro-molecular Solutions" held by the American Chemical Society at its 182nd national meeting at New York City, August 23-28, 1981. The final manuscripts were collected at the meeting so that the authors could expand on the brief accounts that appeared in the Preprints of the Organic Coatings and Plastics Division of the Society.

When planning this symposium we invited experts in the application of the solubility parameter, such as Dr. Charles M. Hansen, Drs. J. E. G. Lipson and J. E. Guillet, and A. J. Tortorello.

Also included were those concerned with the application of solubility theories such as Dr. G. Pezzin, and scientists who are concerned with the kinetics of the dissolution of polymers and the thermodynamic properties of polymeric solutions, such as Drs. A. C. Quano, C. W. Frank and M. L. Huggins. Several timely topics were discussed by C. E. Carraher, M. G. Wyzgoski and I. A. Abu-Isa. C. L. McCormick discussed his recent findings on the Dissolution and Derivatization of Cellulose, and T. C. Shen discussed his more recent membrane work. A special part of the Symposium were D. E. Gregonis, M. E. Lewellyn, and B. M. Moudgil's diverse presentations on hydrophilic polymers.

In addition to the papers presented in New York, additional contributions were submitted and accepted from E. L. Johnson, R. F. Miller, and R. S. Porter.

In our opinion, the chapters in this book, which are written by experts in the field, should prove to be informative to the reader and provide a pattern for additional research. It is a pleasure to note that several authors of chapters in our previous book also provided chapters for this book. These authors are Drs. Charles E. Carraher, Earl J. Johnson, A. C. Ouano and Thomas C. Shen. We wish to express our sincere appreciation to these contributors and to all others who helped to make the symposium successful and for their contribution to this book.

Raymond B. Seymour

G. Allan Stahl

SOME ANSWERED AND UNANSWERED QUESTIONS ABOUT THE SOLUBILITY PARAMETER

by

C. M. Hansen
Scandinavian Paint and Printing Ink Research Inst.
Agern Alle 3, DK 297o
Horsholm, Denmark

ABSTRACT

The solubility parameter has found use in many fields of research and practice, primarily because of its unique predictive capabilities. Other widely used energy parameters such as the Chi parameter and the critical surface tension can be understood better, and perhaps be used more efficiently, if viewed in terms of the more general energy approach offered by the solubility parameter. A still more general theory would seem possible.

KEY WORDS

Solubility parameter; contact angles; adhesion; Chi parameter; surface energy.

INTRODUCTION

The solubility parameter (Hildebrand, 1916; Scatchard, 1931; Hildebrand and Scott, 1950) is an exceptionally useful tool. Like a work of art it can be viewed or used on many levels of activity or intellectual endeavor. The lowest of these is to simply select solvents or mixtures of solvents in an optimum manner. The highest is still to be found.

The broad use the solubility parameter has found in many varied industrial situations demonstrates that it is not a colossus with clay feet. The degree of sophistication of its more practical use varies from one to two to three to even five parameter systems (Karger, Snyder, and Eon, 1978), and the purpetrators of these various schemes seem satisfied with their own results, seeing no need to use anyone else's system, improved accuracy or whatever notwithstanding. This could lead to some confusion, but in fact speaks strongly for the sound fundamental approach the fathers of the concept developed. The fact that the energies involved can and must be considered individually, i.e., multiparameter systems are a necessity for full understanding of what is going on, means that polymers or surfaces, for example, require multiparameter characterization. Single value solubility parameters are ambiguous. This is clear.

Based on the industrial successes we know that the theory is basically correct in the choice of the two parameters molar volume and solubility parameter

1

(cohesive energy density). The theory points out how they, in principle, affect solubility relations. We know equally well that heats of mixing can be both endothermic and exothermic, that is, that the present theory is incapable of accurate predictions. Entropy is, of course, the major problem, but I have not seen any real resolution of the situation. A continuation of the efforts of Patterson (1873) with the solubility parameter offers hope. We still rely largely on empirical correlations and congratulate ourselves on success after success, while the real reason for these successes remains to be found.

Solubility parameter plots are misnomers. We use solubility parameters, which are related to heats of mixing, as axes, and plot solubility regions whose boundaries are really dictated by free energy change being zero. We include the entropy without so stating. We have no generalized theory which can, with suitable simplications, be reduced to practice. Quite to the contrary, the practical use question is largely solved and the theoretical work remains unfinished. This is indeed unfortunate since the practical applications to solubility in general, surface phenomena, inorganic materials, biological materials, etc. clearly point out that what we have now is essentially correct. Systematic studies of mixtures of water with various organic materials are lacking, however.

Examples of the Use of the Solubility Parameter

Barton (1975) has referenced many applications of the solubility parameter in his extensive review article.

    Activity coefficients
    Aerosol formulation
    Chromatography
    Coal solvent extraction
    Compressed gases
    Cosmetics
    Cryogenic solvents
    Dispersion
    Dyes
    Emulsions
    Gas-Liquid solubility
    Grease removal
    Membrane permeability
    Paint film appearance
    Pharmaceutical
    Pigments
    Plasticizers, polymers, resins
    Plasticization
    Polymer and plasticizer compatibility
    Printing ink
    Reaction rate of radical polymerization
    Resistance of plastics to solvents
    Rubber blends
    Solid surface characterization
    Solvent extraction
    Solvent formulation
    Surface tension
    Urea-water solutions

Vaporization of plasticizers
Viscosity of polymer systems
Water-based polymer systems

To this list I would like to add a few more which have been particularly
interesting to me:

Hemodialysis membranes (Chawla and Chang, 1975)
Assymetric membrane formation (Klein and Smith, 1972)
Work of adhesion for liquids on mercury (Beerbower, 1972)
The Rehbinder effect-crushing strength of Al_2O_3 under various liquids
  (Beerbower, 1972)
The Joffe effect-effect of liquid immersion on fracture strength of
  soda-lime glass (Beerbower, 1972)
Friction on polyethylene (treated with $H_2S_2O_7$) (Beerbower, 1972)
Rheology of concentrated polymer solutions (Bagda, 1978)

Recent Surface Applications

I feel the use of solubility parameters, or more broadly, energy parameters,
in the area of contact angles, Θ , and wettability has been neglected. We
know that both surface wetting and surface dewetting can be characterized
using conventional solubility parameter plots (Hansen, 1970, 1972a, 1972b,
1976, 1978, Hansen and Pierce, 1974). See Figure 1. This type of plot
confirms the need for multiparameter surface characterizations, with at
least three parameters (D, P, H).   There is complete analogy between and
good agreement with the plot of Cos Θ versus liquid surface tension first
suggested by Zisman (1964) and a similar plot using Cos Θ and distances to the
center of interaction on a "solubility parameter" plot (like Figure 1) instead
of liquid surface tension. In general, the further removed from the boundary
of the region of interaction a liquid is, the greater the contact angle.
We are currently studying these effects, both for advancing and receding
contact angles, the latter being more relevant to coatings than the former
and also giving "essentially" linear plots on Cos Θ versus liquid surface
tension plots (Hansen 1976, Hansen, 1979). The consequences of the above
would be that the slope of a surface tension plot is related to the angle
of incidence of the respective liquid series to the given region of
interaction. If the liquids in the probe are all series within a relatively
short distance from the boundary, all contact angles will be low and the
surface tension plot will have low slope. If higher energy liquids are
included, higher contact angles will also be found and the surface tension
plot will be steeper. We are collecting data to clarify this and other
aspects of contact angle plots (such as the absolute value of the critical
surface tension) at the present time, anticipating that this oversimplified
view may require some modification. Full details will first be published
when the results of this study are analyzed.

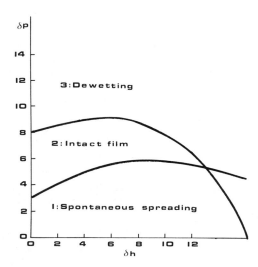

FIGURE 1: Solubility Parameter Interpretation of Surface Energetics for Poly-
           ethylene; a Positive Approach.

           Region 1 includes those liquids which spontaneously spread when
           applied as droplets, i.e., a positive effect.

           Regions 2 and 3 include those liquids which give a contact angle
           when a droplet is applied, i.e., a negative effect, commonly used
           to characterize surfaces. An intact film remains if a film is
           applied (ASTM D-2578-67) in Region 2, i.e., a positive approach.

           Region 3 includes those liquids which retract from the surface when
           applied (ASTM-D-2578-67) as a film in a self-contained de-adhesion
           test. This negative effect is commonly used to describe the "Wetting
           Tension" of the surface.

Potential Future Advances

I feel that advanced computer techniques such as UNIFAC (Abrans and Prausnitz, 1975; Fredenslund, Gmehling, and Rasmussen, 1977; Oiski and Prausnitz, 1978) will improve predictive ability for solubility relations still more for those with access to such computers. Whether or not to consider it the "theory" or not, is still an open question, but the advent of UNIFAC has caused our institute to once more take up research related to the solubility parameter from a more fundamental point of view, after many years of relatively passive activity. The basic idea is that if one can calculate activity coefficients with reasonable accuracy, one should be able to calculate phase relations. I also feel signals in the work of Beerbower (1971, 1972) have been overlooked, at least in a recent review (Barton, 1975). This is for the future. For the present, I think we have more capability in terms of energy characterizations than many recognize, the above mentioned empiricism notwithstanding.

Energy Spectroscopy by Liquid Analysis (ESLA)

When I was approached to prepare a paper for this symposium, my first reaction was to call it "Energy Spectroscopy by Liquid Analysis - (ESLA)". Testing this title on a few individuals quickly led to its rejection, since none of those I asked knew what I was referring to.

The meaning of the phrase is, that given a sufficient number of systematically chosen energy probes of known energy values (liquids of known energy characteristics i.e., cohesive energy densities i.e., solubility parameters or, alternatively, surface energies), one can construct an energy response spectrum for any material which interacts with the energies encompassed by it. Those having response energies outside this range will, of course, not interact. The response can be solubility, swelling, adsorption, contact angles, dewetting, spontaneous spreading, changes in mechanical effects (or who knows what - let us not limit our creativity). These variations in physical effects allow energy spectroscopy by liquid analysis of the material in question in terms of the response to the energy spectrum. Some would prefer to call this procedure a solubility parameter (or contact angle) study, and while this is often acceptable, it is much too restrictive a term for what is really being done. One can use the results of such a study on level one and, for example, select alternate solvents for a system. See Figure 2. On level two it allows estimation of the potential for compatibility of two polymeric materials with each other, for example (Hansen, 1967). See Figure 3. On level three one might estimate the nature of polymer adsorption on (pigment) surface where both polymer (or specific groups on it) and surface have been subjected to energy analysis. See Figure 4. Further examples could be given on each of these levels.

The levels of application and effects mentioned above are, of course, macroscopic manifestations of molecular interactions. Perhaps molecular orbital calculations will some day help to clarify some of these molecular interactions on a still another level. At any rate, I like Burrell many years ago in his (the first) Borden Award paper (Burrell, 1968), feel there is far more to be discovered relative to the solubility parameter than can be imagined today. Burrell after exposing the shaky foundations and lack of predictive capability of the very extensively researched Flory-Huggins Chi parameter compares it with the solubility parameter writing:

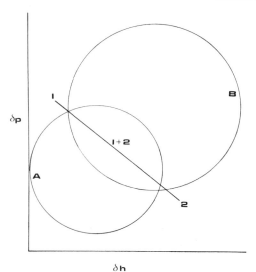

Figure 2: Solvent Selection. Miscibility can be predicted in complicated
           systems such as pictured here where a mixture of two non-solvents
           (1+2) produced a clear solution with two polymers (A+B) (Hansen
           1967). No alternative exists to this unique predictive ability.

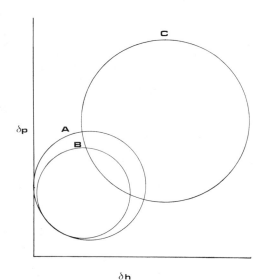

Figure 3: Polymer Miscibility. Polymer miscibility requires energetic
           similarity such as polymers A and B, but this is not necessarily
           sufficient to ensure miscibility if one has a high molecular weight.
           Polymer C lacks energetic similarity to polymers A and B.

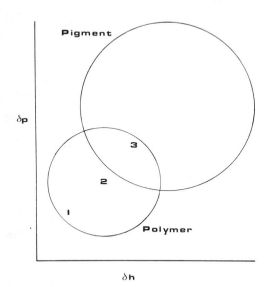

Figure 4: Pigment Dispersion Stability. Solvents, polymers and pigments
can be characterized by solubility parameter. This allows
interpretation of rather complicated dispersion stability
conditions. As pictured, poor solvent 1 will force the polymer
over to the pigment surface. Very good solvent 2 will remove
the polymer from the pigment or extend it greatly into
solution, probably causing instability. Solvent 3 will compete with the polymer for adsorption sites on the pigment
surface. The final result is difficult to judge.

"An appeal is hereby made to our academic fraternity to
abandon the shaky thermodynamic structure which has
been built on false premises and construct a new foundation out of the proven clay of industrial worth.

Burrell continues by quoting a footnote which Hildebrand wrote over thirty
years ago (Hildebrand, 1949). I feel it is still fitting:

"The quantitative limitations set forth are not so
serious as to prevent the theory from being qualitatively very serviceable. We seek qualitative and
relative solubility data far more often than exact
figures. We seek the best or sometimes the poorest
solvent for a certain solute. We seldom want to know
a solubility to, say 1% and, indeed, we seldom control temperature or purity to a corresponding degree.

If we do need a solubility to that accuracy we
must rely upon measurement, better measurements,
indeed, than many in the literature. All theory
can do for us in that case is to select, out of
the scores of solvents upon our shelves, the few
likely to serve our purpose".

Conclusion

Our progress in the last 30 years has been minimal, at most being an improvement of accuracy by adding on (or dividing up) parameters. Better understanding, yes, perhaps, but I feel we have only touched something and have not yet
grasped it. A higher level has not yet been reached, but I am sure it is
there. How to grasp it is the major unanswered question. Some clues may be
found in the breadth of the applications possible, newer, high-power computer
techniques such as UNIFAC, gas chromatographic techniques, the work of
Patterson and coworkers on the entropy problem, or else in seemingly completely
different disciplines such as molecular orbital theory, wetting (adsorption/
adhesion) theory, and acid/base or "charge" theories. Some are on the right
track in isolated disciplines, but a creative combination of the signals seem
necessary to arrive at the ultimate truth of the matter. How best do we predict interactions among materials in general? Do we know enough to put it
all together? If not now, when? And who will do it?

Acknowledgement

This paper received financial support from the Swedish Board for Technical
Development.

References

Abrans, D. S., and J. H. Prauesnitz (1975), J. Am. Inst. Chem. Eng., 21, No. 1,
116.

Bagda, E. (1978), Defazet, 32, No. 10, 372.

Barton, A. F. M. (1975), Chem. Rev. 75, No. 6, 731.

Beebower, A. (1971), J. Colloid Interface Sci., 35, 126.

Beebower, A. (1972), "Boundary Lubrication" Army Research Office Scientific
and Technical Applications Forecast.. Control No. DAHC 19-69-0033, Washington.

Burrell, H. (1968), J. Paint Techn., 40, No. 520, 197.

Chawla, A. S., and T. M. S. Chang (1975), J. Appl. Poly. Sci., 19, 1723.

Fredenslund, A., J. Gmehling, and P. Rasmussen (1977), "Vapor-Liquid
Equilibrium using UNIFAC", Elsevier, Amsterdam.

Hansen, C. M. (1967), Farg och Lack, 13, No. 6., 132.

Hansen, C. M. (1970), J. Paint Techn., 42, No. 550, 660.

Hansen, C. M. (1972a), ibid, 44, No. 570, 57.

Hansen, C. M. (1972b), Chem. Tech., 2, 547.

Hansen, C. M. (1976), Farg och Lack, 22, No. 11, 373.

Hansen, C. M. (1978), XIV FATIPEC Congress Book, p. 97 (Budapest).

Hansen, C. M., and P. E. Pierce (1974), Ind. Eng. Chem. Prod. Res. Dev., 13, 218.

Hansen, W. (1979), "Advances in Printing Science and Technology", p. 288, W. H. Banks, Ed, International Assoc. of Research Institutes for the Graphic Arts Industry, Pantech, London: Plymouth.

Hildebrand, J. H. (1916), J. Am. Chem. Soc., 38, 1452.

Hildebrand, J. H. (1949), Chem. Rev., 44, 37.

Hildebrand, J. H., and R. L. Scott (Second Edition 1936) (Third Edition 1950), "The Solubility of Non-electrolytes", Reinhold.

Karger, B. L., L. R. Snyder, and C. Eon (1978), Anal. Chem., 50, No. 14, 2126.

Klein, E., and J. K. Smith (1972), Ind. Eng. Chem. Prod. Res. Dev., 11, No. 2, 207.

Oiski, T., and J. H. Prausnitz (1978), Ind. Eng. Chem. Proc. Des. Dev., 17, No. 3, 333.

Patterson, D. D. (1973), J. Paint Techn., 45, No. 578, 37.

Scatchard, G. (1931), Chem. Rev., 8, 321.

Zisman, W. A. (1964), "Advances in Chemistry Series Vol. 43", Chapter 1, R. F. Gould Ed. American Chemical Society, Washington, D. C.

APPLICATIONS OF THE SOLUBILITY PARAMETER CONCEPTS
IN POLYMER SCIENCE

Raymond B. Seymour
University of Southern Mississippi
Hattiesburg, MS 39401

Water has served as a solvent for many centuries. Prior to the 19th century,
solutions of bitumens, shellac and fossil resins were produced by dissolving
these naturally-occurring resins in appropriate solvents.

Since only a few organic solvents, such as alcohol, ethyl ether, naphtha, and
turpentine were available, the selection of an appropriate solvent was not
difficult. Solutions of gutta percha, rubber, and cellulose nitrate were made
in the 19th century and while some solvent systems, such as ethanol and ethyl
ether used for making collodion were ingenious, the selection of solvents
continued to be empirical.

Fortunately, acetone, butanol, toluene, methanol, isopropyl alcohol, chlorinated
hydrocarbons, and alkyl acetates were all produced before the early 1930's.
Hence, a wider variety of solvents was available when many of the new synthetic
resins were introduced prior to World War II.

As more new resins and new solvents were introduced, the empirical method
became more cumbersome. In an attempt to quantify the "like dissolves like"
rule of thumb, polymer scientists and technologists derived many empirical
solubility tests. These tests included the kauri-butanol test, the aniline
point, solvent index, the dimethyl sulfate test, the dilution ratio and constant
viscosity procedure.

These empirical tests were based on precipitation or viscosity values at
specific concentrations. The kauri-butanol value was determined by titrating
20g of a 33% solution of kauri resin in butanol with a hydrocarbon solvent
until turbidity was observed. The aniline point was the temperature at which
a mixture of equal volumes of aniline and solvent became turbid.

The solvent index is the ratio of the viscosity of a resin in a standard solvent
to its viscosity in a new solvent. The dimethyl sulfate test indicated the percentage of aromatic and unsaturated hydrocarbons in a hydrocarbon solvent.

The dilution ratio is the ratio of the diluent to true solvent which produces
turbidity when the former is added to a dilute solution of cellulose nitrate.
In the constant viscosity procedure, the viscosity of solutions of nitrocellu-
lose in the solvents and diluent at several different concentrations are plotted

to obtain solutions of constant viscosity.

The solubility parameter concept was developed by Hildebrand and Scott in the early 1920's. However, this important parameter for predicting solubilities was overlooked by many polymer scientists until its usefulness was emphasized by Burrell in 1955.

The use of the solubility parameter concepts which was developed for "regular solvents" was extended to more polar solvents by Crowley and Hansen in the 1960's. The polymer-solvent interaction parameter (μ) developed independently by Huggins and Flory is also very valuable to polymer scientists.

The Hildebrand solubility parameter values (δ) which vary from 7.0 H for pentane to 23.4 H (hildebrand units) for water are a measure of the inter-molecular forces present in the solvent molecules. The δ values vary from about 6H for polytetrafluoroethylene to about 15 H for cellulose. Thus, one could readily predict that pentane and water and cellulose and polytetra-fluoroethylene would be incompatible.

Amorphous polymers are usually soluble in solvents having δ values which are within 1.8H of the polymers. The δ values for aliphatic hydrocarbons increase as the molecular weights of the solvents increase. In contrast, the δ values of polar solvents decrease as their molecular weights increase.

The solubility parameter of volatile liquids may be determined from the energy of evaporation (ΔH), molecular weight (M) and density (D) as shown by the following equation in which R is the gas constant and T is the Kelvin tempera-ture.

$$\delta = \left[\frac{D(\Delta H - RT)}{M} \right]^{1/2}$$

The solubility parameter of nonvolatile solids may be determined from Small's relationship as shown below. The G values in this equation are molar attraction constants. The molecular weight of the repeat unit is used for M when calculating δ values for polymers.

$$\delta = \frac{D \Sigma G}{M}$$

Gardon has shown a relationship between δ values and the polarizability of solvent molecules. Thus, the solubility parameter is related to molecular structure, boiling point, flash point, density, and index of refraction. The relationship of log δ to boiling point (T_b), density (D), and molecular weight (M) is shown by the following equations:

$$\log \delta \approx 0.50 \log \frac{T_b D}{M} + 0.635 \text{ for non polar liquids}$$

$$\log \delta \approx 0.40 \log \frac{T_b D}{M} + 0.665 \text{ for moderately bonded liquids}$$

$$\log \delta \approx \log \frac{T_b D}{M} \text{ for hydrogen bonded liquids}$$

The use of aqueous urea solutions, for the swelling of cellulose and proteins
was the result of many years of empirical investigations. However, the general
knowledge of solubility parameters was used to develop coatings in polymerizable
solvents, such as polybutyl methacrylate in furfuryl alcohol and oil soluble
polycyclohexylstyrene.

The solubility parameter concept was also used to develop the use of copolymers
of ethylene and vinyl acetate as pour point depressants in distillate fuels.
The copolymer compositions were selected so that their solubility parameter
values were within 1.8 H of that of the oil. The solubility parameter of the
copolymers were calculated by use of Small's equation and determined by titration of toluene solutions with poor solvents and from the maximum viscosity
of solutions of the copolymers in solvents with known δ values.

The solubility parameter concept has also been used successfully to produce
useful compositions in coextruded pipe, for the production of useful plasticizers for the selection of low profile resins which are soluble in the polyester
prepolymer but insoluble in the polymer and for the selection of solvents for
fibers, such as polyacrylonitrile.

The solubility parameter has also been used successfully for the selection of
solvents for cellulose and chitin. It had been determined empirically that
cellulose was soluble in aqueous solutions of mineral acids, zinc chloride,
calcium thiocyanate, lithium halides, alkaline solutions, quatenary ammonium
hydroxide, copper, cobalt, nickel, zinc and cadmium ammonia hydroxide, amines,
dimethyl sulfoxide (DMSO) and methylamine, bis (β,γ-dihydroxypropyl)disulfide
and amine oxides.

Solubility parameter concepts were used to select DMSO and dimethylacetamide
(DMAc) for use in solvent systems for cellulose and chitin. Chitin is
soluble in DMAc containing 5 percent lithium chloride and cellulose is soluble
in this solvent system, in dimethylformamide (DMF) in the presence of chloral
and pyridine and in DMSO in the presence of formaldehyde (usually added as
paraformaldehyde).

The solubility parameter concept has also been used to select solvents for
the heterogeneous solution polymerization of vinyl monomers. Thus, rapid
heterogeneous copolymerization of styrene and maleic anhydride takes place in
benzene, toluene, xylene, and cumene which have δ values which differ from
the copolymer by at least 1.8H. The copolymerization is fastest in benzene
in which the difference (Δδ) is 1.8H and slowest in cumene in which Δδ is
2.5H.

The solubility parameter concept was also used to produce insoluble polymers
of ethyl acrylate in cyclohexane, styrene in hexane, vinyl acetate in decane
and methyl methacrylate in hexane or 1-propanol. It is interesting to note
that the polymethyl methacrylate precipitated as formed in both a less polar
and a more polar solvent since in both instances the Δδ values were greater
than 1.8H.

Block copolymers have been prepared by the addition of another monomer to the
precipitated polymer which is a "trapped free radical" or macroradical. The
new monomer molecules must diffuse into the precipitated macroradical and this
process can occur if the Δδ values between monomer and macroradical are less
than 3.2H.

Thus, styrene, styrene-maleic anhydride, methyl methacrylate and acrylonitrile may all form block copolymers with styrene-maleic anhydride copolymers. However, since the polystyrene domains are soluble in benzene, termination occurs when the polystyrene domain exceeds 30% of the total block by weight. The polystyrene block becomes increasingly larger as the solvent is changed to toluene, xylene and cumene.

There is no apparent limit to the length of the polyacrylonitrile block. However, styrene will not add to the acrylonitrile block copolymer. Yet, because of the formation of a charge transfer complex, equimolar mixtures of styrene and maleic-anhydride or styrene and acrylonitrile will add to this block and styrene will add to the new block.

Blocks of acrylic acid, acrylonitrile, ethyl methacrylate, styrene, vinyl acetate, vinylpyrrolidone and styrene-co-acrylonitrile were produced by adding the appropriate monomers to methyl methacrylate macroradicals in hexane or 1-propanol.

It has been recognized that polymers which are insoluble in their monomers will precipitate, as formed, and continue to grow at a rapid rate. This phenomenon has been observed for vinyl chloride and acrylonitrile but not for methacrylonitrile because the $\Delta\delta$ values for monomer and polymer in the latter case is less than 1.8H.

Block copolymers may be formed by adding vinyl monomers to vinyl chloride or acrylonitrile macroradicals providing the $\Delta\delta$ between the monomer and the macroradical is less than 3.2H. However, styrene ($\delta=9.2H$) will not diffuse into the acrylonitrile macroradical ($\delta\approx13.0H$) and hence, block copolymers cannot be prepared by using this sequence. Nevertheless, because they form charge transfer complexes with δ values of 11.0H and 11.9H, respectively, equimolar mixtures of styrene and maleic-anhydride or styrene and acrylonitrile will form block copolymers with acrylonitrile macroradicals.

Most important, block copolymers of acrylonitrile and styrene can be produced by adding acrylonitrile to styrene macroradicals produced in 1-propanol or hexane. In this system, the $\Delta\delta$ between monomer and macroradical is 1.6H. Of course, the block copolymer is the same regardless of whether one adds acrylonitrile to styrene macroradicals or vice versa.

SOLUBILITIES AND SOLUBILITY PARAMETERS OF POLYMERS FROM INVERSE GAS CHROMATOGRAPHY

J. E. G. Lipson and J. E. Guillet

Department of Chemistry, University of Toronto,
Toronto, Canada M5S 1A1

ABSTRACT

Gas chromatography was used to determine the weight fraction solubilities of small organic compounds at infinite dilution in solid polychloroprene, poly(butadiene–acrylonitrile), poly-(ethylene–vinylacetate), and cis-1,4-polybutadiene. The infinite dilution solubility parameters (δ_2^∞) for these polymers were also calculated. Using sets of polar and nonpolar probes was found to have no effect on the values obtained for δ_2^∞. This experimental approach is especially suited to the application of Flory–Huggins theory, since the polymer is the concentrated phase, and the infinite dilution of the probe means there will be little volume change on mixing. The experimental data were also used to calculate weight fraction activity coefficients, partial molar heats of mixing, and heats of solution for each probe over the experimental temperature range, as well as Flory–Huggins χ parameters. The validity of dividing χ into χ_H and χ_S was investigated by comparing experimental results with the predictions of Flory–Huggins and Regular Solution theory.

KEYWORDS

Solubility; solubility parameters; polymers; inverse gas chromatography; Flory–Huggins χ parameter; activity coefficients; heat of solution; Regular Solution theory.

INTRODUCTION

In the past decade gas chromatography (GC) has developed into a useful tool for studying solid polymers. This technique overcomes some of the experimental difficulties associated with solution work, and yields information for a system where the polymer, not the small molecule, makes up the bulk of the solution. The technique of using a GC with a solid polymer stationary phase and injecting small organic probes was first reported by Smidsrød and Guillet (1969); since then, studies have included determination of percent crystallinity in polymers (Hudec, 1977; Braun and Guillet, 1975a), calculation of surface areas (Braun and Guillet, 1975b; Galin and Rupprecht, 1978), glass transition work (Braun and

Guillet, 1976; Schneider and Călugăru, 1975; Ito and coworkers, 1978), diffusion of organic vapor into solid polymer film (Gray and Guillet, 1974), and calculation of thermodynamic data for the probes mixing with the solid polymer (DiPaola-Baranyi, Braun and
Guillet, 1978; Patterson and Robard, 1978; Brockmeir, McCoy and Meyer, 1972; Liu and
Prausnitz, 1976; DiPaola-Baranyi and Guillet, 1978). The latter area is the subject of
this report, in particular for the case of probe at infinite dilution in the polymer.

The experimental quantity of interest in GC is the specific retention volume V_g

$$V_g = (t_R F/w_L)(760/P_0)J_3^2 \tag{1}$$

t_R is the retention time for the solute (probe) minus that for an inert marker (such as
methane), which accounts for the dead volume in the column. F is the flow rate of the
carrier gas quoted at atmospheric pressure and 273.16 K. w_L is the weight in grams of
the polymer in the column, determined by comparing calcination results for a sample of
coated support with results for uncoated support. $760/P_0$ is used so that V_g will be quoted
at standard pressure, and J_3^2 is a correction factor to account for the finite compressibility of the gas in the column, and is given by (Cruickshank, Windsor and Young, 1966)

$$J_3^2 = \frac{3}{2}\left[\frac{(P_i/P_o)^2 - 1}{(P_i/P_o)^3 - 1}\right] \tag{2}$$

where P_i and P_o are the column inlet and outlet pressures, respectively.

Before the application of gas chromatography to polymer studies, Everett (1965) had derived an expression for the activity coefficient of the solute mixing with the solvent in
terms of V_g and pure component data, including M_2, the molecular weight of the coating
on the stationary phase. To avoid having to estimate polymer molecular weights, Patterson and coworkers (1971) modified this equation, using the weight fraction (instead of mole
fraction) rationalized activity coefficient.

$$\ln(a_1/w_1)^\infty = \ln\left[\frac{273.16\,R}{P_1^0 V_g M_1}\right] - \frac{P_1^0(B_{11} - V_1)}{RT} \tag{3}$$

P_1^0, M_1, B_{11} and V_1 are, respectively, the pure vapor pressure, molecular weight,
second virial coefficient and molar volume of the probe. The second term in eq. (3) arises
out of using an equation of state for the probe which describes a slightly imperfect gas.
The expression for the activity coefficient was originally derived for the case of bulk absorption of solute into the stationary phase, and a linear sorption isotherm for the solute.
The implications of this will be discussed in the experimental section.

Under Henry's Law conditions, which are present, the partial pressure of a solute vapor
(p_1) can be related to its solubility (s_1) in the solvent

$$H_1 = p_1/s_1 \tag{4}$$

where H_1 is the Henry's constant. If $p_1 = 1$ atm, then

$$H_1 = 1/s_1^{(1)} \tag{5}$$

For a solute that can be described as an ideal gas, eq. (3) can be rewritten

$$a_1 p_1^0 / w_1 = 273.16 \, R / V_g M_1$$

But $a_1 p_1^0 = p_1$, and w_1 is just the weight fraction solubility of the probe in the polymer. For a $p_1^0 = 1$ atm

$$s_1^{(1)} = V_g M_1 / 273.16 \, R = 1 / H_1^\infty \tag{6}$$

In this manner weight fraction solubilities (therefore Henry's constants) can easily be calculated using retention data. The method described here is essentially the same as that outlined by Liu and Prausnitz (1976), who included in their expression a correction term to account for the solubility of the carrier gas in the polymer. This correction was found to change the solubilities by less than 1%, and is not included in the tabulation of results for this work.

Recall that

$$\Delta F_{mix} = \Delta H_{mix} - T \Delta S_{mix} \tag{7}$$

ΔS_{mix} is calculated using a lattice model for the solution, and accounts only for the combinatorial entropy effects; it is given by (Flory, 1953; Huggins, 1942)

$$\Delta S_{mix} \simeq (N_1 \ln \phi_1 + N_2 \ln \phi_2) \tag{8}$$

where N_1, N_2 and ϕ_1, ϕ_2 are, respectively, the number of molecules and volume fractions of components 1 and 2. ΔH_{mix} is also based on a lattice calculation and is given by

$$\Delta H_{mix} = R T N_1 \phi_2 \chi \tag{9}$$

where χ, the Flory-Huggins interaction parameter is defined as

$$\chi = \frac{z x \Delta w_{12}}{RT} \tag{10}$$

z is the number of nearest neighbors for a site on the lattice (it can vary), x is the ratio of molar volumes of the polymer and probe and Δw_{12} represents the change in free energy for the quasi-chemical reaction of replacing half of a 1-1 and a 2-2 contact with a 1-2 contact.

$$\Delta w_{12} = w_{12} - \frac{1}{2}(w_{11} + w_{22}) \tag{11}$$

Because Δw_{12} is a free energy change it can be divided up into components:

$$\Delta w = \Delta w_H - T \Delta w_S \tag{12}$$

Therefore χ can also be divided:

$$\chi = \chi_H + \chi_S \tag{13}$$

χ_S is often considered to be a correction term to account for the effects of nonrandom mixing on ΔF_{mix}. A more detailed discussion of χ_S is included with the experimental results.

χ_H is related to ΔH_{mix} and using Regular Solution theory (Hildebrand, Prausnitz and Scott, 1970) it can be shown that

$$\chi_H = (V_1/RT)(\delta_1 - \delta_2)^2 \tag{14}$$

where δ's represent solubility parameters, defined as

$$\delta = (\Delta E_{vap}/V)^{1/2} \tag{15}$$

Through its connection to ΔF_{mix}, χ can also be related to the infinite dilution activity coefficient

$$\chi = \ln(a_1/w_1)^\infty - \ln(v1/v2) - (1 - V_1/\overline{M}_{2n}v2) \tag{16}$$

$\overline{M}_{2n}$, the number average molecular weight, is large enough that the last term in parentheses is effectively equal to one. Using V_g and pure component data, χ parameters are easily calculated.

Solubility parameters have found extensive industrial application, usually in compatibility studies; if eq. (14) is correct, then $\Delta H_{mix} = 0$ for $\delta_1 = \delta_2$, and ΔF_{mix} will reach its minimum value. Polymers are nonvolatile, so there is no way to calculate δ_2 directly. In the past, a popular technique has been the swelling experiment: try different liquids with a range of δ's and for the one that swells the polymer sample the most $\delta_2 = \delta_1$. An alternate approach has been derived using GC data and a combination of Flory–Huggins and Regular Solution theory. Combining eq. (13) and eq. (14) (DiPaola-Baranyi and Guillet, 1978)

$$\chi = (V_1/RT)(\delta_1 - \delta_2)^2 + \chi_S \tag{17}$$

and rearranging

$$(\delta_1^2/RT) - (\chi/V_1) = (2\delta_2^\infty/RT)\delta_1 - [(\delta_2^\infty/RT)^2 + (\chi_S/V_1)]$$

A plot of the left-hand side against δ_1 should give a slope and an intercept containing δ_2^∞ at the experimental temperature T.

EXPERIMENTAL

Solutes were used without further purification. Polychloroprene and cis-1,4-polybutadiene were obtained from Polysciences. Poly(butadiene-acrylonitrile) (Krynac) was generously donated by Dr. J. F. Henderson of Polysar, and poly(ethylene-vinylacetate) was obtained from Aldrich.

To prepare the column material an accurately weighed sample of polymer was dissolved in xylenes with slow heating and constant stirring. A pre-weighed amount of Chromosorb G support, 70/80 mesh, was stirred in, and the solvent was slowly evaporated. After vacuum drying near 50 °C for ca. 72 h, the coated support was sieved through 70/80 mesh and packed into a 3 ft x 0.25 in. o.d. copper tube column using a mechanical vibrator to aid with even packing. The weight of the polymer on the stationary phase (w_I) was determined by calcination using a blank support correction. Column parameters are listed in Table 1. A Hewlett Packard 5840 gas chromatograph with dual flame ionization detector

TABLE 1   Stationary Phase and Column Parameters

Polymer	Source	Polymer mass in column (g)	Percent loading
Polychloroprene	Polysciences	0.5014	4.98
Poly(butadiene-acrylonitrile), 34% w/w acrylonitrile	Polysar	0.4848	4.88
Poly(ethylene-vinylacetate), 40% w/w vinyl acetate	Aldrich	0.3826	4.03
cis-1,4-Poly(butadiene)	Polysciences	0.4848	4.95

was used. Nitrogen was the carrier gas, and flow rates were measured from the end of the column using a soap bubble flowmeter, and quoted at 273.16 K and 760 atm. From a Hamilton gas-tight syringe, small amounts of probe vapor (ca. 1.2 μl) were injected, along with 0.2 μl of $CH_4(g)$ to act as a marker for the dead volume in the column. Net retention times were taken as time of probe peak maximum minus time of CH_4 peak maximum. The inlet pressure and the pressure drop across the column were read via a mercury manometer (± 0.05 Torr). The experimental temperature range was 65-85 °C. All work was carried out at least 50 ° above T_g to ensure that bulk absorption was occurring.

P_0 is the sum of P_{atm} and the pressure drop across the column. Solute vapor pressures were found (over the experimental temperature range) using literature constants (A, B, C) for the Antoine equation

$$p_1^0 = A + [B/(t + C)]$$ (18)

where t is in °C. Second virial coefficients for the probes were computed using

$$B_{11}/V_c = 0.430 - 0.886(T_c/T) - 0.694(T_c/T)^2 - 0.0375(n-1)(T_c/T)^{4.5}$$ (19)

T_c and V_c represent the critical temperature and volume of the probe, and n represents the number of carbon atoms in the corresponding linear alkane.

Using eq. (15) and assuming that the solute could be described as an ideal gas, solute solubility parameters were found at experimental temperatures:

$$\delta_1 = [(\Delta H_{vap} - RT)/V_1]^{1/2}$$ (20)

where

$$\Delta H_{vap} = \Delta \bar{H}_1^\infty - \Delta H_{sol}$$ (21)

ΔH_1^∞ is the partial molar enthalpy of the probe at infinite dilution in the solution.

$$\Delta \bar{H}_1^\infty = R \frac{\partial \ln(a_1/w_1)}{\partial(1/T)}$$ (22)

and ΔH_{sol} is the enthalpy of solution.

$$\Delta H_{sol} = -R \left[\frac{\partial \ln V_g}{\partial (1/T)} \right] \tag{23}$$

Three assumptions were made concerning the experimental procedure.

1. Bulk absorption represents the major part of the interaction. This was tested by making up a second column for each polymer with a different percent loading (different amount of polymer per amount support), therefore changing the surface to bulk ratio of the coating. The retention volume would change for a probe with a significant amount of surface retention, making it necessary to extrapolate V_g to the value it would have for a column with infinite percent loading.

2. The probe has a linear sorption isotherm, i.e., the concentration of the probe in the gas phase is linearly related to its concentration in the bulk phase. This was tested by making sure that the retention time (concentration profile) was independent of the sample size. For a probe whose retention time did change, an extrapolation to zero sample size would be needed.

3. An equilibrium is present between the bulk and the moving phase. This means that the probe has sufficient time to diffuse through the bulk of the polymer. This was tested by varying the flow rate and making sure that the retention volume did not change. A change in V_g would necessitate an extrapolation to zero flow rate.

Assumptions 1 and 2 arise because eq. (3) was derived for conditions of bulk absorption and a linear sorption isotherm for the probe. None of the extrapolations listed were necessary for the experiments described here.

RESULTS AND DISCUSSION

Weight fraction solubilities for various probes in the four polymers are quoted in Table 2 at 75°C, the midpoint of the experimental temperature range. Flory-Huggins χ parameters are also quoted at 75°C in Table 3.

According to Hansen (1967) it is possible to divide up the solubility parameter into contributions arising from dispersion interactions, polar (dipole and induced dipole) interactions and hydrogen bonding (or charge transfer) interactions.

$$\delta^2 = \delta_d^2 + \delta_p^2 + \delta_h^2 \tag{24}$$

If this is correct, then it seems likely that using nonpolar probes would contribute to δ_2^∞ only through δ_d^2, resulting in a δ_2^∞ smaller than that which would be found using probes capable of other interactions. It should therefore be possible to calculate the different contributions to δ_2^∞ by using different sets of probes. With this in mind, solutes were chosen for each polymer such that roughly half had some dipole interaction or hydrogen bonding possibilities. Plots were done (eq. 17) for the polar, nonpolar and combined sets of probes and δ_2^∞ was calculated for each case. Figure 1 shows the plots. It was almost impossible to distinguish the least squares fit for the different sets of probes, and only the fit which includes all of the probes is shown. The linear relationship described by eq. (17) is obeyed very well, supporting the method of analysis used. Results for the

TABLE 2   Weight Fraction Solubilities for Various Polymers at 75 °C

Solute	Poly-chloroprene	Poly(butadiene-acrylonitrile)	Poly(ethylene-vinylacetate)	cis-1,4-Polybutadiene
n–Octane	0.344	0.423	0.802	1.04
1–Chlorobutane	0.200	0.306	0.308	0.313
Carbon tetrachloride	0.375	0.503	--	0.686
1–Octene	0.391	0.338	0.807	0.992
n–Decane	0.355	2.63	4.90	--
Benzene	0.274	0.432	0.358	0.366
Chloroform	0.283	0.591	0.561	0.415
Methyl propyl ketone	0.504	0.758	0.510	--
n–Butanol	0.466	--	0.759	--
Toluene	--	1.10	0.973	1.06
n–Butylcyclohexane	--	3.69	--	--
Chlorobenzene	--	--	2.78	--
n–Hexane	--	--	0.117	--
n–Pentane	--	--	0.040	0.053

TABLE 3 χ Parameters for Various Polymers at 75 °C

Solute	Poly-chloroprene	Poly(butadiene-acrylonitrile)	Poly(ethylene-vinylacetate)	Polybutadiene
n–Octane	1.15	1.30	0.562	0.424
1–Chlorobutane	0.375	0.298	0.206	0.313
Carbon tetrachloride	0.281	0.330	--	0.065
1–Octene	0.883	1.38	0.418	0.336
n–Decane	1.25	1.49	0.578	--
Benzene	0.106	−0.004	0.098	0.204
Chloroform	0.054	−0.344	0.375	0.052
Methyl propyl ketone	0.144	0.080	0.392	--
n–Butanol	0.641	--	0.871	--
Toluene	--	0.013	0.053	0.100
n – Butylcyclohexane	--	1.14	--	--
Chlorobenzene	--	--	−0.057	--
n–Hexane	--	--	0.578	--
n–Pentane	--	--	0.632	0.516

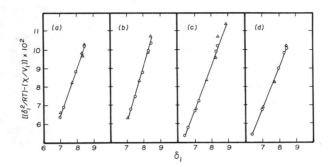

Fig. 1. Infinite dilution solubility parameters for various polymers at 75 °C using experimentally determined χ parameters: ($\triangle$) polar probes, ($\bigcirc$) nonpolar probes. (a) δ_2^{∞} (slope) = 8.83 ± 0.22, (b) δ_2^{∞} (slope) = 10.0 ± 0.29, (c) δ_2^{∞} (slope) = 8.26 ± 0.17, (d) δ_2^{∞} (slope) = 7.90 ± 0.14.

δ_2^{∞}'s using polar/nonpolar/all probes are given in Table 4 along with the appropriate literature range for δ_2.

δ_2^{∞} values generally fell within the estimated range for δ_2 although the exact relationship between the two is not clear. As the data show, there appeared to be effectively no difference (within experimental error) between the δ_2^{∞}'s obtained using the different sets of probes.

TABLE 4 Comparison of δ_2^{∞} (slope) at 75 °C from Polar/Nonpolar Probes

Polymer	δ_2^{∞} (all)	δ_2^{∞} (polar)	δ_2^{∞} (nonpolar)	δ_2 (25°) lit.
Poly(ethylene-vinylacetate)	8.26 ± 0.17	8.26 ± 0.38	8.13 ± 0.10	$8.44 - 9.67$
cis-1,4-Poly-butadiene	7.90 ± 0.14	8.12 ± 0.44	7.83 ± 0.11	$7.16 - 8.60$
Poly(butadiene-acrylonitrile)	10.0 ± 0.29	10.3 ± 0.60	9.73 ± 0.28	$9.84 - 10.18$
Polychloroprene	8.83 ± 0.22	8.58 ± 0.39	9.08 ± 0.18	$8.18 - 9.38$

One explanation for these results was outlined in an earlier work (Lipson and Guillet, 1981). The dipole interaction energy was calculated using the dipole moments of the most "polar" polymer and probe combination (polychloroprene with methyl propyl ketone),   assuming an angle between them of 180° and a separation distance made up of the van der Waals radii for the end atoms. This energy contribution was estimated to be -1.782 kcal/mol at 75°C, compared to kT at the same temperature of 0.69 kcal/mol. This was for the strongest dipole interaction under ideal conditions; it is therefore possible that many of the dipole contributions to the interaction are swamped out by kT. The energy contribution from hydrogen bonding pairs was estimated to be of similar magnitude, and there are even less hydrogen bonding combinations possible. It would appear that to see significant contributions from δ_p and δ_h, extremely polar pairs would be needed, making the concept of the three dimensional solubility parameter not very useful for typical inverse GC experiments.

The consistency of the different δ_2^∞'s can be explained in another way. χ_S is supposed to make up for the inadequacies of a combinatorial-only entropy of mixing, by accounting for nonrandom mixing effects. These can arise out of preferential orientations in the mixture, and/or from volume differences between the two components. In the linear plots done, χ_S is part of the intercept, while δ_2^∞ is calculated using the slope. This means that δ_2^∞ (slope) would be relatively insensitive to the kind of effects embodied in χ_S, therefore would not be dependent on the type of probes used, which is what the experimental results showed.

The two interpretations are connected in the sense that χ_S is often treated as a correction to the entropy of mixing, and from the rough calculations done, it is expected that this correction would be small unless very polar components are used.   Inverse GC therefore provides an easy method for the determination of weight fraction solubilities and infinite dilution solubility parameters; in addition, the infinite dilution of the probe makes some of the theoretical approximations used in treating the data more reasonable than they would be for solution studies, in particular, the assumption (used in the original derivations of Flory-Huggins theory, and in Regular Solution theory) that $\Delta V_{mix} = 0$.

Next it was decided to look at χ_S and its contribution to χ. Using $\chi_{(total)}$ from the experimental results, and assuming that χ_H could be described by eq. (14), χ_S and its percent contribution to χ was calculated. Results were similar for all four polymers and only the data for poly(ethylene-vinylacetate) are summarized in Table 5, which also lists χ. For the four polymers the majority of polymer-probe combinations showed that χ_S was larger than 50% of χ; far more than a correction. In some cases, χ_S was even greater than χ.

Recall that χ_S arises out of Guggenheim's (1948) work on deriving the energy of mixing using a lattice model (as Flory-Huggins theory does). Originally he defined the difference in total intermolecular potential Δu_{12}, due to the formation of a 1-2 contact at the expense of half of a 1-1 and a 2-2 contact (eq. 11). He then equated Δu_{12} to the free energy change for this quasi-chemical reaction, and relabelled it Δw_{12}. Being a free energy change Δw_{12} could then be divided (eq. 12), leading to the division of χ into χ_H and χ_S. This is where χ_S comes from. Its physical interpretation remains unclear; although it is usually considered to be a correction term, in practice it is often taken as a constant between 0.3 and 0.4. As χ typically lies between 0-1, this would make χ_S more than just a correction, which is borne out by the results from this work. As χ_S is supposedly linked to ordering effects and volume differences, it seemed reasonable to plot $|\chi_S/\chi|$ against δ_1 to see how the percent contribution changed as the probe changed. The plot for poly(ethylene-vinyl acetate) is shown in Fig. 2. The other three polymers were similar in that they showed

TABLE 5   Contribution of χ_S to χ for
Poly(ethylene-vinyl acetate)

Probe	$|\chi_S/\chi|$ x 100	χ_{total}
1-Chlorobutane	76 %	0.21
1-Octene	11	0.42
n-Butanol	13	0.87
Chlorobenzene	209	-0.06
Chloroform	112	-0.38
Methyl propyl ketone	100	0.39
Toluene	99	0.05
n-Decane	44	0.58
n-Hexane	2	0.58
Benzene	95	0.10
n-Octane	27	0.56
n-Pentane	4	0.63

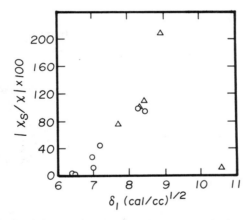

Fig. 2.   Change in contribution of χ_S to χ with changing δ_1.
○ nonpolar probes, △ polar probes.

the same trend of $|\chi_S/\chi|$ increasing with δ_1, although not necessarily in a linear fashion. In an effort to determine whether ordering in the system or volume differences were contributing the greatest amount to the interaction, $|\chi_S/\chi|$ was plotted separately as a function of the energy and the volume part of δ_1, but there was no discernible correlation in either of the plots. It would seem that either χ_S is genuinely becoming more important as δ_1 increases, and/or that χ_H, as derived from Regular Solution theory, is becoming a less accurate reflection of the enthalpic interaction as δ_1 increases, leaving everything not included in χ_H to be mopped up by χ_S. In looking at the data, the first possibility seems the less likely of the two; why should a combination like cis-1,4-polybutadiene and toluene have $|\chi_S/\chi|$ of 74% ?

The expression for χ_H arises out of the derivation for ΔE_{mix} ($= \Delta H_{mix}$) for a regular solution (Hildebrand and Wood, 1933). The major assumptions in this derivation are that the intermolecular potential (ϕ) between any two molecules can be described as a simple attractive potential of the form $-k/r^6$, and that the form of the probability distribution function $W(r)$ for both components arises out of a thermally disturbed close-packed arrangement for spherical molecules, the $W(r)$'s for the two components being simply related.

The energy of mixing is calculated by finding the difference in total intermolecular potential between the solution and the pure components. This is done through integration of $W(r) \times \phi$. Leaving out the repulsive part of ϕ may not be that critical; it should only become important at small r, and if $W(r)$ is large enough ϕ would be swamped out. The second assumption seems to be the more questionable of the two for a polymer-small molecule mixture, where the two components may pack in a very different fashion, giving rise to $W(r)$'s that cannot be easily related. The above two assumptions make up what is known as the "geometric mean approximation":

$$k_{12} = [(r_{11} + r_{22})^3 / 8 r_{11}^3 r_{22}^3] (k_{11} k_{22})^{1/2} \tag{25}$$

or, effectively

$$k_{12} \approx (k_{11} k_{22})^{1/2}$$

For r_{11}, r_{22} differing by as much as 50%, the constant in front is still only 1.06. It is the geometric mean assumption which tidies up the equation for ΔE_{mix}, enabling Hildebrand to define his solubility parameter δ. Improving the level of approximation could be possible, by improving upon the form of the potential function or, even more importantly, changing the form of the probability distributions, but this would invalidate the geometric mean approximation, resulting in a messier calculation which would not yield solubility parameters.

CONCLUSIONS

The experimental conditions used in Inverse Gas Chromatography are better suited to the application of Flory-Huggins and Regular Solution theory than the conditions in solution studies, where the small molecule is in the bulk phase. In addition, the method of analysis used in this work predicts a linear relationship (eq. 17) which is confirmed by the experimental results. Solubilities, solubility parameters and other data are easily obtained using this approach.

However, there is an inconsistency in the physical picture of what is occurring between polymer and probe as described by the theory and as revealed by the experimental results. Although variables used in comparison studies, such as δ, are useful for semi-quantitative work, it can be misleading to divide up parameters into components and assign a physical interpretation to them. From the results presented here it would seem that χ is such a variable. Division of χ provides little insight into the polymer-probe interaction, and it may be that a lattice model is not the best framework for examining concentrated polymer solutions.

ACKNOWLEDGEMENTS

The authors would like to thank Dr. S. G. Whittington for invaluable discussions. In addition, the authors are grateful to the Natural Sciences and Engineering Research Council of Canada for financial support of this research. Scholarship support (to J.E.G.L.), through a University of Toronto Open Fellowship and an Ontario Graduate Scholarship, is also gratefully acknowledged.

REFERENCES

Braun, J.-M., and J. E. Guillet (1975a). J. Polym. Sci., Polym. Chem. Ed., 13, 1119.
Braun, J.-M., and J. E. Guillet (1975b). Macromolecules, 8, 882.
Braun, J.-M., and J. E. Guillet (1976). Macromolecules, 9, 340.
Brockmeir, N. F., R. W. McCoy, and J. A. Meyer (1972). Macromolecules, 5, 464.
Cruickshank, A. J. B., M. L. Windsor, and C. L. Young (1966). Proc. R. Soc. (London), A271, 295.
DiPaola-Baranyi, G., J.-M. Braun, and J. E. Guillet (1978). Macromolecules, 11, 224.
DiPaola-Baranyi, G., and J. E. Guillet (1978). Macromolecules, 11, 228.
Everett, D. H. (1965). Trans. Farad. Soc., 1637.
Flory, P. J. (1953). Principles of Polymer Chemistry, Cornell University Press, Ithaca.
Galin, M., and M. C. Rupprecht (1978). Polymer, 19, 506.
Gray, D. G., and J. E. Guillet (1974). Macromolecules, 7, 244.
Guggenheim, E. A. (1948). Farad. Soc. Trans., 44, 1007.
Hansen, C. M. (1967). J. Paint Technol., 39 (505), 104.
Hildebrand, J. H., J. M. Prausnitz, and R. L. Scott (1970). Regular and Related Solutions, Van Nostrand Reinhold Co., New York
Hildebrand, J. H., and S. E. Wood (1933). J. Chem. Phys., 1, 817.
Hudec, P. (1977). Makro. Chem., 178, 1187.
Huggins, M. L. (1942). J. Phys. Chem., 46, 151.
Ito, K., H. Sakakura, K. Isogai, and Y. Yamashita (1978). J. Polym. Sci., Polym. Lett. Ed., 16, 21.
Lipson, J. E. G., and J. E. Guillet (1981). J. Polym. Sci., Polym. Phys. Ed., 19, 1199.
Liu, D. D., and J. M. Prausnitz (1976). Ind. Eng. Chem. Fundam., 15, 330.
Patterson, D., and A. Robard (1978). Macromolecules, 11, 640. 690
Patterson, D., Y. B. Tewari, H. P. Schreiber, and J. E. Guillet (1971). Macromolecules, 4, 356.
Schneider, I. A., and E.-M. Călugăru (1975). Eur. Polym. J., 11, 857.
Smidsrød, O., and J. E. Guillet (1969). Macromolecules, 2, 272.

THE SOLUBILITY PARAMETER CONCEPT IN THE DESIGN
OF POLYMERS FOR HIGH PERFORMANCE COATINGS

Anthony J. Tortorello, Mary A. Kinsella and Richard T. Gearon

DeSoto, Inc., 1700 S. Mt. Prospect Rd.,
Des Plaines, IL   60018

ABSTRACT

Coatings for aircraft application are required to meet a variety of severe conditions including flexibility over a broad temperature range and resistance to fluids
covering the spectrum of polarity from hydrocarbons to water. Upon consideration
of all the various film performance requirements, fluid resistance is considered
to be the most crucial.

The design of synthetic resins to resist the test fluids began with the characterization of the solubility parameter value for each fluid. A spectrum of fluid
solubility parameter values was then constructed and a region corresponding to
resistance to the entire body was identified. Using the method of atomic group
contributions, acrylic copolymers having the desired solubility parameter value
were synthesized. Both anionic and cationic aqueous dispersions were prepared and
evaluated for clear-film performance. The performance was as expected for all
fluids except water.

KEYWORDS

Aircraft coatings; anionic dispersion; cationic dispersion; fluid resistance; solubility parameter; aqueous coatings

INTRODUCTION

Prior to the enactment of Rule 66 by Los Angeles County interest in water-based
coatings for industrial use was limited primarily to automotive finishes.[1] Since
then, additional legislative restrictions[2] and a changing perspective regarding
the supply and economic advantages of solvents have led to the onset of technology
change.

In the refinishing of its service aircraft, the U.S. Air Force is governed by the
same restrictions which apply to private industry. Recognizing the needs of a
changing technology and compliance with federal emission guidelines, the Air Force
has continuously sponsored research to replace its solvent-based epoxy-polyamide
primer and urethane topcoat since the early 1970's. This sponsorship has taken
the form of several contracts ranging in scope from direct emulsification of
primer and topcoat components[3] to development of high solids coatings[4] to development of water-based coatings.[5]

The following report presents in part the results of Air Force contract F33615-78-C-5096. The study proposes the development of an aqueous resin system intended to function as the pigment binding vehicle for an aircraft topcoat or primer.

Table 1 summarizes the military specification which characterizes the applied film properties of the solvent-based urethane topcoat. The requirements of the primer are similar. And any prospective replacement must display this performance.

TABLE 1

MILITARY SPECIFICATION MIL-C-83286

COATING, URETHANE, ALIPHATIC ISOCYANATES
FOR AEROSPACE APPLICATIONS

5% Salt Spray	No blistering, cracking, corrosion, or loss of adhesion after 500 hours of exposure.
100% Relative Humidity	No blistering, cracking, softening, or loss of adhesion after 720 hours of exposure.
Accelerated Weathering	After 500 hour exposure the coating should exhibit 60% impact flexibility, no more than 10% loss of original gloss, and no color change.
Fluid Resistance	A decrease of no more than one pencil hardness unit after immersion in water (4 days, 100°F), lubricating oil (24 hours, 250°F), hydrocarbon fluid (7 days, room temperature), and hydraulic-fluid (7 days, room temperature). A decrease of no more than two pencil hardness units after immersion in Skydrol 500B fluid (7 days, room temperature).
Film Flexibility	No cracking, crazing, or loss of adhesion of coating when elongated 60% by impacting mandrel.
Low Temperature Flexibility	No cracking or loss of adhesion when bent around 3/8 in (9.5 mm) diameter cylindrical mandrel after four hours at -65°F (-54°C). (Test immediately after removal from cold box).
High Temperature Resistance	No loss of adhesion or flexibility after four hours at 300°F (149°C).
60° Gloss	>90

The results of a preliminary screening of the commercial marketplace for aqueous resins clearly showed that successful performance could not be predicted by generic polymer classification.[5] The study then proceeded to design of synthetically novel polymers, dispersion into aqueous medium, formulation with film aids, and comparison of clear-film (unpigmented) performance to the established criteria.

DISCUSSION

Upon reviewing all the application requirements the two most challenging appear to be flexibility and fluid resistance. Since chemical resistance is usually achieved by crosslinking in the applied film and since crosslinking is usually accompanied by a deterioration of mechanical flexibility, a paradox seems to result. Furthermore, the Flory-Huggins equation[6] (Eq. 1) predicts that cross-linking is not the sole answer to achieving chemical resistance. Even network

$$q_m^{5/3} \simeq (V_o/\nu_e) \; (\tfrac{1}{2}-\chi_1)/V_1 \qquad\qquad \text{(Eq. 1)}$$

structures are appreciably swollen by solvents having a favorable solvent-polymer interaction parameter (χ_1).

Given the inverse relationship between chemical resistance and mechanical flexi-bility any resin found to display optimum chemical resistance in the non-cross-linked state could achieve the final resistance requirements with minimal cross-linking and hence optimum flexibility. The problem simplifies to one of achieving optimum chemical resistance in the linear polymer.

Fluid Resistance and the Solubility Parameter Concept

Fluid resistance may be interpreted as a measure of incompatibility between a polymer and a fluid which can be viewed as a solvent. In considering the design of a polymer to maximize resistance to a particular fluid, the solvent-polymer interaction parameter (χ_1) is difficult to apply. Alternately, the compatibility of a solvent and a polymer has been expressed in terms of the solubility parameter concept.[7]

Hildebrand[8] has shown that for a solution process to occur the solubility param-eter value of the solvent must be nearly equal to that of the solute. Conversely, incompatibility is predicted when there is a disparity between the two values. Hence, the design of novel polymers for enhanced fluid resistance can be guided by a broad distinction between the solubility parameter value of the polymer and that of the fluid. Furthermore, this enhanced performance should display no de-pendence on generic polymer classification. Acrylics should perform as well as urethanes, etc.

Acrylic Copolymer Design

The solubility parameter value has been defined according to equation 2 as the

$$\delta = (\frac{\Delta E}{Vm})^{\tfrac{1}{2}} \qquad\qquad \text{(Eq. 2)}$$

square root of the cohesive energy density i.e. the ratio of energy of vaporiza-tion and molar volume. This definition renders the solubility parameter concept more suitable to novel polymer design than the solvent-polymer interaction param-eter. Small[9] and later Rheineck and Lin[10] showed that the contribution of each atomic grouping comprising the molecular structure could be totalled to estimate the solubility parameter value. More recently, Fedors[11] has provided an exten-sion to the number of groups available.

Solubility Parameter Calculation for Acrylic Monomers. According to Fedors' method the contribution of each atomic group to the molar energy of vaporization and volume is summed over the molecular structure and the square root of the ratio is taken as the solubility parameter value. For n-butyl acrylate the calculation is as follows:

$$\{CH_2-CH\}$$
$$CO_2(CH_2)_3CH_3$$

Group	Number	Δe	Δei	Δv	Δv_i
CH_3	1	1125	1125	33.5	33.5
CH_2	4	1180	4720	16.1	64.4
CH	1	820	820	-1.0	-1.0
CO_2	1	4300	4300	18.0	18.0
			10965		114.9

$$\delta = \left(\frac{\Sigma \Delta ei}{\Sigma \Delta v_i}\right)^{\frac{1}{2}} = \left(\frac{10965}{114.9}\right)^{\frac{1}{2}} = 9.77$$

Table 2 presents the values of the summation of Δe_i and Δv_i terms for some acrylic monomers. These values will be required for the calculation of the solubility parameter of copolymers.

Table 2

Energies of Vaporization and Molar Volumes of Some Acrylic Monomers

Monomer	$\Sigma \Delta ei$ (cal/mol)	$\Sigma \Delta vi$ (cm^3/mol)
n-Butyl acrylate (BA)	10965	114.9
2-Ethylhexyl acrylate (2EHA)	15270	179.6
Ethyl acrylate (EA)	8605	82.7
Methyl acrylate (MA)	7425	66.6
Vinyl acetate (VAc)	7425	66.6
Methyl methacrylate (MMA)	8080	81.9
Styrene (Sty)	9630	86.5
Acrylonitrile (AN)	8100	39.1
Acrylic acid (AA)	8600	43.6
Itaconic acid (Ita)	15910	70.0
Acrylamide (AM)	12000	32.6
Methacrylamide (MAM)	12655	47.9
2-Hydroxyethyl acrylate (HEA)	15780	75.3
2-Hydroxyethyl methacrylate (HEMA)	16435	90.6

Solubility Parameter Calculation for Acrylic Copolymers. In considering the design of copolymers having a preferred solubility parameter value, the entire molecular structure must be viewed. For acrylic copolymers the molecular structure of the repeating unit is ethylene with varying mole fractions of functionality pendant to the ethylene backbone. The solubility parameter of the copolymer then becomes a function of summation of the energy of vaporization and molar volume terms for each monomer comprising the polymer multiplied by the mole fraction of that monomer. The following example illustrates the calculation:

Monomer	Mol.Frac.(X)	$\Sigma\Delta ei$	$\Sigma\Delta ei*X$	$\Sigma\Delta v_i$	$\Sigma\Delta v_i*X$
EA	0.6173	8605	5311.9	82.7	51.1
Sty	0.3043	9630	2930.4	86.5	26.3
AN	0.0562	8100	455.2	39.1	2.2
Ita	0.0222	15910	353.2	70.0	1.6
			9050.7		81.2

$$\delta = \left(\frac{\Sigma\Delta ei*X}{\Sigma\Delta v_i*X}\right)^{\frac{1}{2}} = \left(\frac{9050.7}{81.2}\right)^{\frac{1}{2}} = 10.56$$

RESULTS

Two series of acrylic copolymers were taken from hypothetical design, through
aqueous dispersion, to clear-film evaluation in order to verify the applicability
of the solubility parameter concept in the design of polymers for high performance coatings.

Aircraft Test Fluids

The design of a polymer for enhanced fluid resistance according to the solubility
parameter concept must first begin with a characterization of the solubility parameter value of the fluid. Table 1 identifies five fluids which an aircraft
coating may contact. These materials are: TT-S-735 type III hydrocarbon (T-3);
diester lubricating oil (LO); Mil H-5606 hydraulic fluid (H-F); Skydrol 500B hydraulic fluid; and water.

Characterization of the solubility parameter for each fluid was accomplished by
boiling point[8] and surface tension[12] techniques. Where possible a comparison to
literature reported values[13,14] was made. A spectrum of fluid solubility parameters was then constructed and is displayed in figure 1.

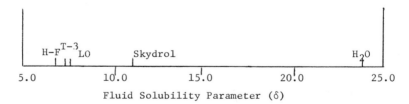

Figure 1. Solubility parameter spectrum of
erosive aircraft fluids.

The most striking feature of this spectrum is the extreme gap between Skydrol and
water. A resin designed to resist the body of fluids as a whole should have a
solubility parameter value falling approximately midway within this gap.

Anionic Acrylic Aqueous Dispersions

A series of acrylic solution polymers was designed varying solely in solubility
parameter. Part of the design included carboxylic acid monomers. These carboxy-
functional polymers when reacted with volatile amines and mixed with water give
rise to colloidal dispersions stabilized by the carboxylate anion. Upon film
formation the amine evaporates along with other volatile components leaving the
free carboxylic acid for subsequent crosslinking chemistry.

Implicit to the other applied film requirements is the specification that performance must be achieved under ambient conditions. One of the few reactions of
carboxylic acids known to occur at room temperature is addition to the aziridine
ring.[15] The reaction when applied to crosslinking utilizes polyfunctional
aziridine resins as illustrated in equation 3.

$$(Eq. 3)$$

Anionic Dispersion Properties. The series of synthetic acrylic dispersions prepared varied in solubility parameter value from 11 to 14. Attempts to maintain
consistent molecular weight and functionality level were made to eliminate any
potential effects of these variables on performance. Table 3 summarizes the
properties of these anionic acrylic formulations.

Table 3
Anionic Acrylic Formulation Properties

Monomer Comp.	2722-37	2722-54	2722-70	2722-75	2722-75a	2722-82
	BA/MA/AN/AA	2EHA/Sty/ AN/AA	EA/HEA/ AN/AA	AM/HEA/ AN/AA	AM/HEA/ AN/AA	BA/VAc/AN/AA
Solubility param.(δ)	11.4	12.0	12.8	13.8	13.8	11.4
Acid value	24.4	21.9	21.9	24.6	24.6	31.9
Tg C (est)	28	34	57	59	59	41
Aziridine resin*	XAMA-7	XAMA-7	XAMA-7	XAMA-7	NONE	XAMA-7
Solids, percent	25.0	31.2	39.5	22.5	18.2	29.9

*Polyvinyl chemical Industries

The formulations were calculated to have stoichiometric amounts of carboxyl and
aziridine functionalities. This was to insure consumption of the acid which, if
not completely reacted, would provide a site of water sensitivity. Formulation
2722-75a is the only member not utilizing a crosslinking agent. As a curiosity,
this formulation was prepared to examine the need for crosslinking in the case of
specialty resins designed for enhanced chemical resistance.

Formulations 2722-37 and 2722-82 represent an interesting comparison. The polymers are identical with one exception: vinyl acetate (VAc) is substituted directly for methyl acrylate (MA). Since the two monomers are isomeric in atomic
structure, group contributions to the total solubility parameter result in the
same value for each polymer. The significance lies in the fact that comparative chemical resistance should be achieved with the cost advantage associated
with vinyl acetate.

Fluid Resistance Performance. The formulations in Table 3 were spray applied
to treated aluminum substrate, allowed to dry for seven days at ambient temperature and humidity and immersed in each test fluid. Fluid resistance was measured in terms of pencil hardness rating. Each coating was given a rating before
and after immersion. In order from hardest to softest, the pencils used are:
4H, 3H, 2H, H, F, HB, B, 2B, 3B, 4B. Table 4 describes the fluid resistance per-

formance:

Table 4

Fluid Resistance of Anionic Acrylic Coatings

Formulation	Film Thickness,mil	Original Hardness	Lubricating Oil(δ,8)	Water (δ,23)	H5606 (δ,7)	Skydrol 500B(δ,11)	TT-S-735 (δ, 7.5)
2722-37	1.4-2.4	B,HB	HB	<4B	HB	3B,4B	B
2722-54	0.5-2.0	HB	HB,F	3B,<4B	HB	HB	HB
2722-70	1.1-2.2	HB	F	<4B	HB,F	F	F
2722-75	0.8-1.0	HB	H,F	<4B	HB,F	HB	HB
2722-75a	1.0-1.3	HB	F	DF*	HB	HB	HB
2722-82	1.3-2.1	HB	HB	<4B	HB	HB	HB

*Film was dissolved by fluid.

Some entries show two values because all experiments were performed in duplicate. Where different, both values were reported. There appears to be a break-off point in solubility parameter where resistance to Skydrol can be predicted. The break-off occurs somewhere around 12 as evidenced by the performance of 2722-37 (δ = 11.4) and 2722-54 (δ = 12.0). All the resins above solubility parameter 12 display no softening when immersed in Skydrol (or any of the organic fluids). This behavior is consistent with the solubility parameter concept if the value for Skydrol is taken to be as observed around 11 and the others between 7 and 9.

Perhaps the most striking support of this theory is indicated in the performance of 2722-75a. Recall that this formulation was prepared without a crosslinking agent. This resin as a lacquer (not crosslinked) displays resistance to all the organic fluids.

Also worthy of note is the comparative performance of 2722-37 and 2722-82. Both resins have identical solubility parameters but 82 is prepared from less expensive starting materials. The equivalent performance is in agreement with the solubility parameter concept. In fact, 2722-82 displays better resistance to Skydrol. However, formulation 2722-82 was prepared from a resin having a larger acid value than that in 2722-37. The improved Skydrol resistance is believed to be an artifact of increased crosslink density.

The table indicates that all formulations are softened by water. In some cases such as 2722-75a, this is to be expected. But in most other cases water softening is unaccountable. This anomalous moisture sensitivity leads to the suspicion of incomplete cosolvent evaporation or incomplete reaction of carboxyl groups under ambient film formation.

Correlation of Observed Fluid Resistance with Expected Performance. The results of performance listed in Table 4 can be summarized graphically for ready comparison to expected behavior. Figure 2 is an idealized curve relating resistance to a particular fluid as a function of polymeric solubility parameter. As the resin solubility parameter value approaches that of the fluid, softening is expected.

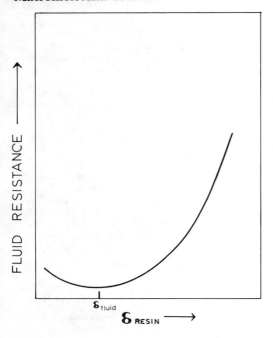

Figure 2. Idealized fluid resistance
as a function of polymeric solubility
parameter.

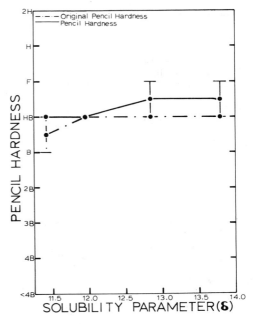

Figure 3. Polymeric solubility parameter effect on resistance to H5606
hydraulic fluid ($\delta \sim 7$).

Figures 3 through 7 summarize the results of Table 4. Each figure represents pencil hardness after immersion in the indicated fluid as a function of resin solubility parameter. Comparison of this curve to the curve of pencil hardness before fluid immersion indicates the degree of attach by (or resistance to) the fluid of interest.

In general, performance is as predicted by solubility parameter theory. Water appears to be the only fluid which does not display the expected behavior. However, as discussed previously, water may be a special case. Effects other than solubility alone may be more significant in predicting resistance to this fluid.

Cationic Acrylic Aqueous Dispersions

In order to test the generality of the concept, a series of polymers requiring different crosslinking chemistry was designed. An alternate reaction known to occur at room temperature is the addition of an amine to an epoxide. Amine functionality can be introduced into a polymeric backbone. When reacted with volatile acids, the resultant ammonium cation can stabilize a colloidal dispersion. Upon film formation, the volatile acid evaporates leaving the free amine for crosslinking chemistry. The crosslinking of polyamines with polyepoxides is shown in equation 4.

$$\text{P-NH}_2 + \text{CH}_2\text{-CH-R-CH-CH}_2 \longrightarrow \text{P-NHCH}_2\text{CH-R-CHCH}_2\text{NH-P} \qquad \text{(Eq. 4)}$$

Since the amino group is to react with an epoxy group, primary or secondary amines are preferrable to tertiary amines. These latter merely catalyze epoxy-epoxy reactions.

Amino-functional Acrylic Copolymers. Acrylic copolymers were prepared using glycidyl methacrylate (GMA) to introduce epoxy functionality. The epoxide was then converted to amine functionality by addition of a ketimine blocked adduct. This adduct was the reaction product of N-methyl-1,3-propanediamine and methyl isobutyl ketone according to equation 5.

$$\text{CH}_3\text{NH(CH}_2)_3\text{NH}_2 + \text{CH}_3\text{CCH}_2\text{CH(CH}_3)_2 \rightleftharpoons \text{CH}_3\text{NH(CH}_2)_3\text{N=C}\begin{smallmatrix}\text{CH}_3\\\text{CH}_2\text{CH(CH}_3)_2\end{smallmatrix} + \qquad \text{(Eq. 5)}$$

$$\begin{smallmatrix}\text{CH}_3\\\text{CH}_3\text{N}\end{smallmatrix}\diagup\begin{smallmatrix}\text{CH}_2\text{CH(CH}_3)_2\\\text{N-H}\end{smallmatrix}$$

After addition to the epoxy polymer followed by hydrolysis upon aqueous dispersion, the cyclic isomer would give secondary amine functionality while the linear isomer gives primary amine functionality. Carbon and proton NMR spectroscopy can identify only the linear isomer as the isolated reaction product.

The sequence of amine addition to the epoxy polymer followed by ketimine hydrolysis is outlined in equations 6 and 7.

$$\text{P-CH}_2\text{CH-CH}_2 + \text{CH}_3\text{NH(CH}_2)_3\text{N=C}\begin{smallmatrix}\text{CH}_3\\\text{CH}_2\text{CH(CH}_3)_3\end{smallmatrix} \longrightarrow \text{P-CH}_2\text{CHCH}_2\text{N(CH}_2)_3\text{N=C}\begin{smallmatrix}\text{CH}_3\\\text{CH}_2\text{CH(CH}_3)_2\end{smallmatrix}$$

$$\text{(Eq. 6)}$$

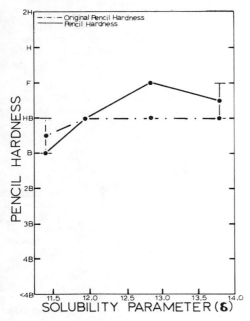

Figure 4. Resin solubility parameter effect on resistance to TT-S-735 Type III hydrocarbon (δ~7.5).

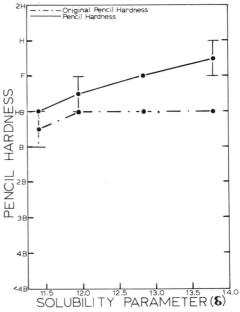

Figure 5. Polymer solubility parameter effect on lubricating oil (δ~8) resistance.

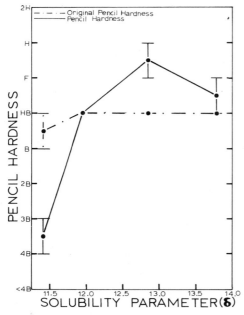

Figure 6. Polymer solubility parameter effect on Skydrol 500B (δ~11)
resistance.

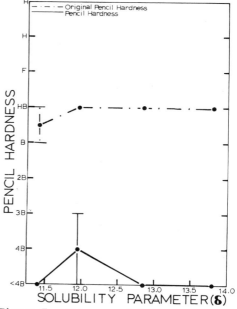

Figure 7. Polymer solubility parameter effect on resistance to deionized water (δ~23).

$$\underset{OH}{\overset{\substack{OH \qquad CH_3}}{\oplus\text{-}CH_2CHCH_2N\text{---}(CH_2)_3N=C}}\overset{CH_3}{\underset{CH_2CH(CH_3)_2}{\diagup}} \;+\; H_2O \longrightarrow \oplus\text{-}CH_2\overset{OH}{C}HCH\overset{H\,\oplus}{\overset{\substack{CH_3}}{N}}\text{---}(CH_2)_3NH_2 \;+$$

$$\overset{O}{\overset{\|}{CH_3CCH_2CH(CH_3)_2}}$$

(Eq. 7)

Cationic Dispersion Properties. The resultant cationic dispersions generated by the previous sequence were blended with polyfunctional epoxies for crosslinking according to equation 4. The series of synthetic cationic dispersions varied in solubility parameter from 9.5 to 12.5. Attempts were made to maintain molecular weight and functionality levels consistent with those of the anionic acrylic series. Table 5 summarizes the properties of these cationic acrylic formulations.

Table 5
Cationic Acrylic Formulation Properties

	2830-06	2722-198	2830-19	2830-13
Monomer comp.	BA/MMA/GMA	MA/AN/GMA	BA/AN/GMA	EA/AN/GMA
Solubility param.(δ)	9.7	11.2	12.0	12.4
Amine eq.wt.	1076.5	1047.6	871.6	899.0
Tg ^{o}C (est)	30	30	34	59
Epoxy resin*	DER 331	DER 331	DER 331	DER 331
Solids, percent	10.2	18.4	9.6	14.7

*Dow Chemical Co.

The values listed for amine equivalent weight do not appear to be consistent with those of the anionic series. However, upon inspection of the amino-acrylic structure two amine groups are found to be present. Since one of these groups is tertiary, it will not participate in direct crosslinking reactions. Thus when adjusted for only the reactive amine, the equivalent weights are consistent.

Fluid Resistance of Cationic Acrylic Formulations and Correlation. As was accomplished for the series of anionic acrylics, curves relating fluid resistance in terms of pencil hardness rating to resin solubility parameter were constructed. Comparison of the curves before and after immersion in each fluid can be used to predict the solubility parameter at which resistance is expected. Figures 8 through 12 illustrate such curves.

Figure 8 depicts resistance to H5606 hydraulic fluid. An unexpected softening is displayed at 12.4 yet the attack is not very severe (a decrease of about one unit).

Figure 9 indicates expected resistance to TT-S-735 hydrocarbon at around 11. And this curve is generally in agreement with the idealized curve in figure 2.

Resistance to diester lubricating oil can be expected for polymers of solubility parameter above 10.5 as indicated in figure 10.

Figure 11 relating resistance to Skydrol 500B displays the same unexpected softening at 12.4 as figure 8 for hydraulic fluid. Other than this anomaly, the curve is similar to ideality and predicts resistance for resins above solubility parameter 11.

Figure 12 indicates that water severely attacks all polymers in the series. As was true for the anionic acrylic series, the case of water evidently does not apply in an assessment of this type.

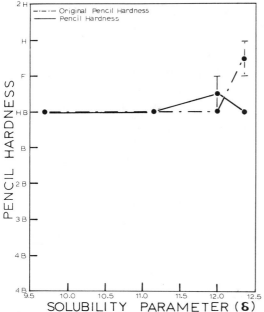

Figure 8.   Effect of cationic acrylic
solubility parameter on resistance to
H5606 hydraulic fluid ($\delta \sim 7$).

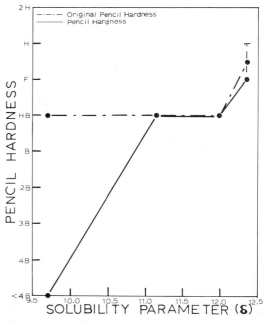

Figure 9.   Effect of cationic acrylic
solubility parameter on resistance to
TT-S-735 type III hydrocarbon ($\delta \sim 7.5$).

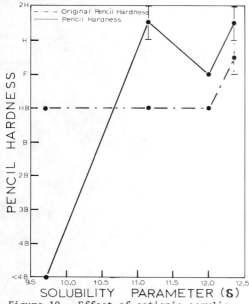

Figure 10. Effect of cationic acrylic solubility parameter on resistance to diester lubricating oil ($\delta \sim 8$).

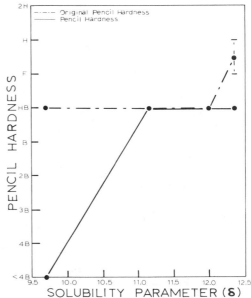

Figure 11. Effect of cationic acrylic solubility parameter on Skydrol 500B ($\delta \sim 11$) resistance.

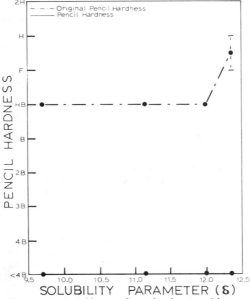

Figure 12. Effect of cationic acrylic solubility parameter on water ($\delta \sim 23$) resistance.

SUMMARY

Using the method of atomic group contributions polymers having preferred solubility parameter values have been synthesized. A series of carboxy-functional
polymers were prepared, dispersed into aqueous medium, and formulated as anionic
water-based coatings. Similarly, a series of amine-functional polymers were
evaluated as cationic aqueous coatings.

Results indicate that the solubility parameter concept in useful in predicting
polymeric resistance to a variety of organic fluids. However, water is apparently a special case. Effects other than solubility may be more significant in
predicting resistance to this fluid.

REFERENCES

1.   Levinson, S.B. J. Paint Technol. 1972, 44(369), 41.

2.   South Coast Air Quality Management District; Proposed Rule 1124, April 13,
     1979.

3.   Vanderhoff, J.W. et al. AFML-TR-74-208 1974.

4.a. McKay, G.L. AFML-TR-79-4041 1979.
  b. McKay, G.L.; Chattopadhyay, A.K. AFWAL-TR-80-4073 1980.

5.a. Tortorello, A.J.; Gearon, R.T. AFML-TR-79-4192 1979.
  b. Tortorello, A.J.; Kinsella, M.A.: Gearon, R.T. AFWAL-TR-80-4197 1981.

6.   Flory, P.J. "Principles of Polymer Chemistry;" Cornell Univ. Press: Ithaca,
     N.Y., 1953; pp. 577-580.

7.a. Burrell, H. Off. Dig. Fed. Paint Varn. Prod. Clubs 1955, 27, 726.
  b. Barton, A.F.M. Chem. Rev. 1975, 75(6), 731.

8.   Hildebrand, J.H.; Scott, R.L. "The Solubility of Nonelectrolytes;" 3d ed.,
     Reinhold Pub. Co.: New York, 1950.

9.   Small, P.S. J. Appl. Chem. 1953, 3, 75.

10. Rheineck, A.E.; Lin. K.F. J. Paint Technol. 1968, 40(527), 611.

11.a. Fedors, R.F. Polym. Eng. Sci. 1974, 14(2), 147.
   b. Fedors, R.F. Polym. Eng. Sci. 1974, 14(6), 472.

12. Burrell, H. in "Polymer Handbook," 2d ed., Brandrup, J.; Immergut, E.H. eds.;
     John Wiley & Sons, Inc.: New York, 1975; p. IV-337.

13. Hoy, K.L. J. Paint Technol. 1970, 42(541), 76.

14.a. Hansen, C.M. J. Paint Technol. 1967, 39(505), 104.
   b. Hansen, C.M.; Skaarup, K. J. Paint Technol. 1967, 39(511), 511.

15. For other examples of aziridine chemistry see: Paquette, L.A. "Principles
     of Modern Heterocyclic Chemistry;" W.A. Benjamin, Inc.: Reading, Mass,
     1968; pp. 1-45.

THE ROLE OF SOLUBILITY IN STRESS CRACKING
OF NYLON 6,6

Michael G. Wyzgoski
General Motors Research Laboratories
Polymers Department
Warren, Michigan 48090

ABSTRACT

The effect of liquid type, moisture content and temperature on the stress cracking behavior of semicrystalline nylon 6,6 has been investigated. Dry nylon is highly susceptible to cracking with critical strains as low as one percent in many liquids. Nylon which has been preconditioned in 50% relative humidity or immersed in water shows almost no propensity to stress crack at room temperature. However, moisturized nylon is still susceptible to stress cracking at low temperature (-40°C). The equilibrium weight gains of the liquids in nylon can be represented on a two-dimensional solubility parameter map and also are proportional to the hydrogen bonding solubility parameter alone. However, unlike previous results for glassy noncrystalline polymers the critical strains for cracking of nylon 6,6 cannot be simply related to solubility. This is due to the fact that rapidly sorbed liquids cause homogeneous stress relaxation to occur in competition with crack initiation. Thus, in general, both the solubility and sorption kinetics must be known in order to predict critical strains of nylon 6,6 in organic liquids.

KEYWORDS

Crazing; cracking; critical strain; moisture content; nylon 6,6; organic liquids; solubility; solubility parameters; stress cracking; stress relaxation.

INTRODUCTION

Under the action of a liquid environment nominally ductile polymers can fail in a brittle manner when subjected to relatively low tensile stresses. This phenomenon, known as environmental stress cracking, is particularly perplexing since it often occurs unexpectedly in seemingly inert liquids. The general subject of environmental stress cracking of polymers has recently been reviewed by Kramer (1979). The related topic of stress crazing has been reviewed by Kambour (1973) as well as Rabinowitz and Beardmore (1972). Considerable effort has been directed at developing techniques to predict stress cracking behavior. For glassy amorphous polymers it has been shown that the propensity for liquids to cause cracking is related to the solubility of the liquid in the polymer (Kambour, 1973). Since the measurement of solubility is often difficult or requires long times, solubility parameter models have been employed to estimate solubility, and thus, predict stress cracking behavior. For example, for polycarbonate, the critical strain values for over eighty liquids or liquid mixtures were well described using a two-parameter solubility parameter model which also incorporated a molar volume term (Jacques and Wyzgoski, 1979).

The present study was undertaken in an attempt to extend this approach to a semicrystalline polymer, namely nylon 6,6. Stress cracking of nylon polymers by organic liquids has previously been reported by Weiske (1964) whereas cracking by inorganic salt solutions has been discussed by Dunn and Sansom (1969). It is not clear to what extent different mechanisms are operating for solvent versus salt cracking. In this report the stress cracking of nylon 6,6 in organic liquids is characterized by measuring the critical strains required to produce cracking. The role of solubility is then addressed by comparing the critical strains with the corresponding equilibrium weight gains and solubility parameters of the liquids employed.

EXPERIMENTAL

Materials

A commercially available nylon 6,6 resin (DuPont Zytel 101) was employed in this study. Liquids were selected based upon their capacity for polar and hydrogen bonding interactions. A series of alcohols was also examined to investigate the role of increasing molecular size. Table 1 lists the liquids employed along with appropriate physical constants such as molar volume and solubility parameters. The latter values were taken from the review by Barton (1975).

Sample Preparation

Since injection molded samples have a high degree of orientation and exhibit an unusual surface morphology, compression molded samples were used in this study. For critical strain measurements a thickness of 0.127 cm was used whereas thinner samples of 0.013 cm thickness were used for weight gain experiments. To minimize oxidation during molding the mold was purged and maintained with dry nitrogen gas. Immediately after molding, the samples were stored in desiccators to prevent moisture absorption.

Moisture Absorption

To assess the influence of absorbed water on stress cracking behavior samples were conditioned in a 50% relative humidity or immersed in distilled water. Weight gains were followed to insure equilibrium was attained. A reproducible 50% relative humidity was obtained by suspending samples over a solution containing 39% by weight of glycerine in water.

Critical Strain Measurements

Previously an elliptical bending form was used to measure the critical strain of 0.318 cm thick samples (Wyzgoski and Jacques, 1977). For the thinner samples used in this study the elliptical form provides strains below the range of interest. Therefore constant strain forms were fabricated in which strips 3.17 cm by 1.27 cm were subjected to a three point bending with a span of 2.54 cm. The loading pin was threaded at 15.7 threads/cm (40 threads/inch) using hexagonal head bolts. A preset strain was established by controlling the displacement of the bolt (i.e. one turn = 0.0635 cm). Each form accomodated four separate samples which were set to cover a range of strains. Critical strain was determined by averaging the highest strain which resulted in no cracking and the lowest strain which caused cracking. For example if four samples set at 0.5, 1.0, 1.5 and 2.0% strain showed cracks at only the two higher levels, the critical strain would be 1.25%. Samples were loaded and allowed to relax in a desiccator for five minutes prior to being completely immersed in the test fluid.

Equilibrium Solubility

To approximate the equilibrium solubilities of the liquids in nylon, weight gains were monitored for thin samples (0.013 cm) which were immersed in the test liquid at 50°C. The use of thinner samples and a somewhat elevated temperature was intended to accelerate the rate of sorption. Because of concurrent extraction and the possibility of increased crystallization during absorption this technique is at best an approximation of the true solubility. To more clearly define the sorption kinetics, measurements were also conducted using the thicker samples (0.127 cm) at a temperature of 40°C as well as -40°C.

Dynamic Mechanical Measurements

A Dynamic Mechanical Analyzer module for the DuPont 990 Thermal Analyzer System was used to measure the glass transition temperature (T_G) of nylon samples. The 0.127 cm thick samples were used and temperature was varied at 5°C/min from -150°C to +150°C. Transition temperatures were determined from the resonant frequency temperature data. Reported values represent the onset of the temperature interval where the modulus rapidly decreases due to the glass transition.

Stress Relaxation Measurements

The rate of decrease in stress versus time for samples strained in a three-point bending mode was measured using a flexural relaxometer. This apparatus was previously described (Wyzgoski and Jacques 1977). The sample size and span were selected to duplicate the loading conditions of the constant strain forms. Load versus time was recorded for samples totally immersed in the liquid environment at room temperature.

RESULTS

Critical Strain Measurements

Fig. 1 shows the cracking exhibited by dry nylon 6,6 samples at strain levels of 0.7, 1.4, 2.1, and 2.8% after the samples were immersed in n-butanol for 10 minutes. For this example the critical strain concept is clearly demonstrated with a value of 1.05% indicated for nylon in n-butanol. Table 1 lists the critical strains (ε_c) for nylon 6,6 in the various organic solvents.

The effect of moisture content on the critical strain of nylon 6,6 is shown in Table 2. No cracking is observed for samples which have been fully saturated (8.5% weight gain) in water. Even for samples equilibrated at 50% relative humidity (2.5% weight gain) no cracking is observed. Thus absorbed water has a tremendous effect on the stress cracking susceptibility of nylon 6,6 in organic liquids.

This is further indicated by the rapid increase in critical strain shown in Fig. 2. In this case only the outer surface has absorbed water since more than 150 days are required to equilibrate a sample 0.15 cm thick (DuPont Zytel Design Handbook). If water is desorbed from the partially moisturized sample the

Fig. 1.   Stress cracking fixture showing cracking of nylon
          6,6 test strips exposed to n-butanol at strains
          of 0.7, 1.4, 2.1, and 2.8%.

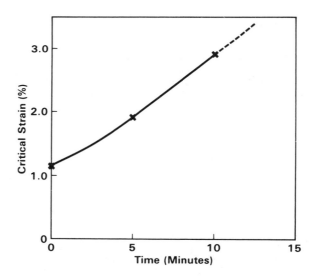

Fig. 2.   Effect of exposure to 50% relative humidity on
          the critical strain of nylon 6,6 in n-butanol.

TABLE 1 Molar Volumes and Solubility Parameters of Selected Organic Solvents in which Critical Strains and Weight Gains of Dry Nylon 6,6 were Measured

Liquid	V_o (ml/mole)	δ_d	δ_ε (δ values are in $cal^{1/2}$ $cm^{-3/2}$)	δ_h	δ_o	Weight Gain (%)	Volume Fraction	ε_c (%)
Low Polarity, Low "h"								
Carbon tetrachloride	97.0	8.7	0	0.3	8.7	2.9	.021	4.0
Toluene	106.8	8.8	0.7	1.0	8.9	3.2	.043	3.7
N-Heptane	147.4	7.5	0	0	7.5	1.0	.017	3.5
Low Polarity, Moderate "h"								
1,4-Dioxane	85.7	9.3	0.9	3.6	10.0	5.5	.061	1.3
Low Polarity, High "h"								
Phenol	87.5	8.8	2.9	7.3	11.8	Solvent for nylon		
m-Cresol	104.7	8.8	2.5	6.3	11.1	Solvent for nylon		
Moderate Polarity, Low "h"								
Methyl isobutyl ketone	125.8	7.5	3.0	2.0	9.3	2.7	.039	1.1
Cyclohexanone	104.0	8.7	3.1	2.5	9.6	7.8	.095	2.1
Acetone	74.0	7.6	5.1	3.4	9.8	3.5	.051	2.2
Moderate Polarity, Mod. "h"								
Diacetone alcohol	124.2	7.7	4.0	5.3	10.2	7.7	.093	1.3
Tetrahydrofuran	81.2	8.2	2.8	3.9	9.5	5.3	.069	2.0
Moderate Polarity, High "h"								
n-Butanol	91.5	7.8	2.8	7.7	11.3	7.5	.106	1.0
n-Propanol	75.1	7.8	3.3	8.5	12.0	11.1	.158	1.1
Ethanol	58.5	7.7	4.3	9.5	13.0	11.1	.161	1.9
High Polarity, Low "h"								
Nitroethane	71.5	7.8	7.6	2.2	11.1	5.0	.055	1.1
Acetonitrile	52.6	7.5	8.8	3.0	12.0	2.6	.038	1.0
High Polarity, Mod. "h"								
Dimethylformamide	77.0	8.5	6.7	5.5	12.1	8.6	.104	2.0
Dimethylsulfoxide	71.3	9.0	8.0	5.0	13.0	16.6	.173	2.1
High Polarity, High "h"								
Ethylene glycol	55.8	8.3	5.4	12.7	16.1	17.9	.184	1.8
Formic acid	37.8	7.0	5.8	8.1	12.2	Solvent for nylon		
Methanol	40.7	7.4	6.0	10.9	14.5	11.8	.171	2.5
Water	18.0	7.6	7.8	20.7	23.4	8.5	.097	1.8

critical strain decreases to the value for a dry sample as shown in Fig. 3. Thus, the effect of water on critical strain is reversible. In addition the beneficial effect of absorbed water is temperature dependent. For example in Table 3 it is noted that at -40°C the critical strain is similar in methyl isobutyl ketone for dry samples and those equilibrated in water. This may be explained in part by the decreased glass transition temperature, Tg, of moisturized nylon 6,6. However, a simple correlation with Tg may not be available. In methonol and n-butanol, "wet" nylon 6,6 samples still exhibit a high critical strain at -40°C; however, samples equilibrated at 50% relative humidity show a low value. Thus the improved resistance to stress cracking of nylon plasticized with water is offset by decreasing temperature in a manner dependent on the specific solvent.

TABLE 2   Effect of Moisture Content on the Critical Strains of Nylon 6,6

Liquid	Sample Conditioning	ε_c (%)
Methanol	Dry	2.5
	50% R.H.	>3.2
	Saturated	>3.2
Butanol	Dry	1.0
	50% R.H.	>3.2
	Saturated	>3.2
Methyl isobutyl ketone	Dry	1.1
	50% R.H.	>3.2
	Saturated	>3.2

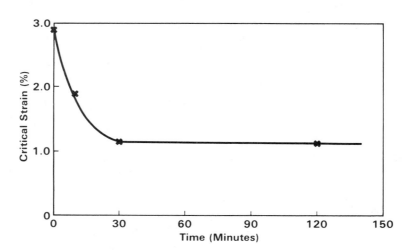

Fig. 3.   Effect of drying (by exposure over $CaSO_4$ dessicant) on critical strain of nylon 6,6.

TABLE 3   Critical Strains of Nylon 6,6 Measured at $-40°C$

Liquid	Sample Conditioning	ε_c(%)
Methanol	Dry	1.0
	50% R.H.	1.0
	Saturated	>2.9
Butanol	Dry	1.0
	50% R.H.	1.1
	Saturated	>3.2
Methyl isobutyl ketone	Dry	1.1
	50% R.H.	1.0
	Saturated	1.0

Weight Gain Measurements

Fig. 4 shows the increase in weight for nylon 6,6 immersed in nitroethane.   For
the thin film an equilibrium level was reached in less than 200 hours.   Slight
decreases in weight gain (less than 1/2%) occurred for longer times probably due
to extraction of low molecular weight species.   Similarly shaped curves were
observed for most of the organic liquids.   In cases where absorption occurred
slowly the maximum weight gain was not as well defined.   For example, Fig. 5 shows
data for nylon 6,6 in methyl isobutyl ketone.   For this liquid, sorption occurred
at a reduced rate with an apparent equilibrium being established only after 3300
hours.   As a worst case example the weight gain in acetonitrile is shown in
Fig. 6.   The nylon film weight initially increased to an apparent equilibrium then
subsequently decreased drastically (Fig. 6).   This behavior is attributed to
recrystallization during sorption as will be discussed later.   In general the
weight gain changes were sufficiently well-behaved (similar to Fig. 4) to permit
an estimate of the maximum liquid absorbed.   These are listed in Table 1.

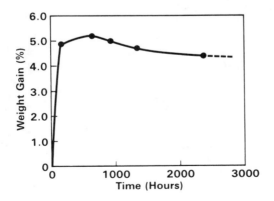

Fig. 4.   Weight increases for nylon 6,6 immersed in
nitroethane at 50°C.

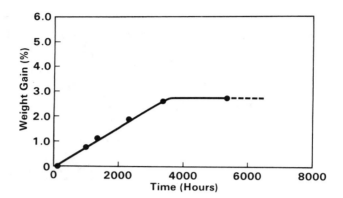

Fig. 5.   Weight increases for nylon 6,6 immersed in methyl
          isobutyl ketone at 50°C.

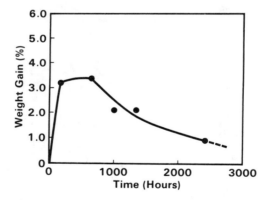

Fig. 6.   Weight increases for nylon 6,6 immersed in
          acetonitrile.

It should be noted that these values serve only as a rough estimate of the polymer-liquid interaction and cannot be treated as equilibrium solubilities. No attempt was made to correct the values for crystalline content of the nylon even though it is presumed that absorption occurred only in the amorphous regions. A similar assumption has been made by previous investigators in the study of nylon-alcohol sorption behavior (Sfirakis and Rogers, 1980) and in the sorption of organic liquids by polyethylene (Rogers, Stannett, and Szwarc, 1959).

DISCUSSION

Correlation of Weight Gains to Solubility Parameters

For predictive purposes as well as general understanding of the crazing phenomenon it was of interest to determine whether solvent interactions with semi-crystalline nylon could be described by a solubility parameter model. Theoretically such a model is permissible only if the crystalline phase is considered inert. That is, the solvent neither dissolves crystalline material nor causes additional crystallization to occur. This possibility was investigated by means of x-ray diffraction and differential scanning calorimetry (DSC) measurements of nylon 6,6 samples which were previously equilibrated in the organic liquids. Only in the case of acetonitrile sorption was a significant change in crystallinity indicated by DSC after 1000 hours immersion. This same sample showed a decrease in solvent uptake at longer times as noted in Fig. 6. For all other liquids no change in crystallinity was evident after solvent swelling.

The crystalline phase could also modify the sorption behavior of nylon by acting to crosslink the amorphous regions. This was assumed by Rogers, Stannett, and Szwarc (1959) in their study of the sorption of organic liquids by polyethylene. This effect does not negate the possible correlation between measured solubility and the solubility parameter but could cause a systematic error in the measured values (proportional to degree of swelling).

In Fig. 7 the measured weight gains for nylon 6,6 are converted to a volume fraction of absorbed liquid and are plotted against the solubility parameters of each liquid. Values given by Hansen in the review by Barton (1975) were employed. A swelling envelope is observed for nylon 6,6 similar to that previously reported for noncrystalline glassy polymers by Kambour (1973). The peak occurs coincident with the reported values of 11.4 to 13.6 $cal^{1/2}$ $cm^{-3/2}$ for the calculated solubility parameter of nylon 6,6 (Askadskii and coworkers, 1977; and Burrell, 1955). The skewed shape of the envelope suggests a strong influence of hydrogen bonding which is typically high for solvents having δ_o greater than 14 $cal^{1/2}$ $cm^{-3/2}$. Also it is noted that a range of solubilities can be exhibited by solvents having a given value of δ_o. Normally such discrepancies indicate that more specific polar or hydrogen bonding interactions are occurring.

Fig. 8 shows how the volume fractions absorbed for each liquid in nylon appear on a solubility parameter map. In this case the total solubility parameter is replaced by the polar and hydrogen bonding components. The ability of the liquid solubility parameters to represent the data in a consistent fashion is considered to be reasonably good. Areas of low, medium and high swelling are arbitrarily drawn, though they cannot be rigorously defined with the available data. Fig. 8 again suggests a strong influence of hydrogen bonding in determining the extent of interaction of a solvent with nylon. For example highly polar liquids with limited hydrogen bonding capacity still exhibit relatively low swelling in nylon.

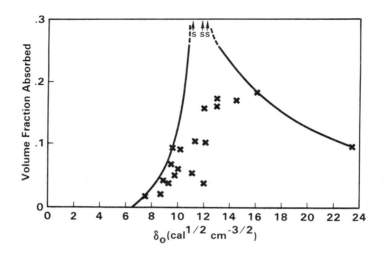

Fig. 7.   Volume fraction of organic liquid absorbed by
           nylon 6,6 versus the corresponding solubility
           parameter of the liquid.

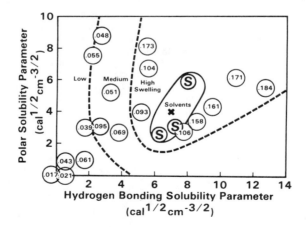

Fig. 8.   Solubility parameter map for absorption of
           organic liquids by nylon 6,6.   Numbers
           represent volume fractions absorbed.

Plots of volume or weight fraction absorbed as a function of δ_h showed considerable scatter. However, when the number of molecules absorbed was plotted as a function of δ_h an excellent correlation was noted as shown in Fig. 9. This relationship serves to emphasize the role of hydrogen bonding and also provides a means for estimating the extent of interaction (predict weight gain a priori) for organic liquids which are only partially soluble in nylon. However some caution must be exercised in using this correlation since its theoretical basis is not clear. For example, liquids which completely dissolve nylon (formic acid, m-cresol, and phenol) have values of δ_h of 6.3 to 8.1 $cal^{1/2}$ $cm^{-3/2}$. The latter would not fit the seemingly linear relationship shown in Fig. 9 since they hypothetically have molar ratios of infinity.

Based upon these results it is concluded that solubility parameters can be used to roughly estimate the interaction expected between nylon 6,6 and organic liquids. This is possible in spite of the semicrystalline nature of the polymer.

<u>Critical Strain - Solubility Relationships.</u> For a variety of glassy polymers Kambour (1973) has demonstrated that the critical strain for crazing is related to the equilibrium solubility of the liquid crazing agent in the polymer. As solubility increases the critical strain decreases. Based upon the previous section a general relation of critical strain to either the measured weight gains or the solubility parameters of the liquids might be expected. In Fig. 10 the critical strains measured for nylon 6,6 are plotted against the corresponding weight gains which are again expressed as volume fractions of liquid absorbed. The solid line represents the type of curve previously reported for glassy polymers. The nylon critical strains initially show a general decrease similar to the relationship for glassy polymers. However there is also an indication of an upturn at higher solubilities (dashed line). Moreover there is a significant scatter in critical strain at any given value of solubility. Thus it appears that by knowing the solubilities of a given liquid in nylon one would still not be able to make an accurate prediction of the corresponding critical strain for the liquid.

The relation between critical strain and the semiempirical solubility parameters of the liquids was also examined. For this purpose it was necessary to specify the solubility parameters of nylon. A value for the total solubility parameter of 12.0 $cal^{1/2}$ $cm^{-3/2}$ was selected for nylon 6,6. This value is an average of the published calculated values and the values for the three liquids which completely dissolve nylon. The overall solubility parameter was further divided into dispersion, polar, and hydrogen bonding components by selecting values based upon the solubility parameter map in Fig. 8. Corresponding to the X in Fig. 8 values of δ_{P_h} = 7.0 $cal^{1/2}$ $cm^{-3/2}$ and δ_{P} = 4.0 $cal^{1/2}$ $cm^{-3/2}$ were selected. This results in a value of δ_{P_d} = 8.9 $cal^{1/2}$ $cm^{-3/2}$ to arrive at a total of 12.0 $cal^{1/2}$ $cm^{-3/2}$.

Since the liquids selected vary primarily in polar and hydrogen bonding potential the critical strain data was plotted on a map having polar and hydrogen bonding solubility parameters as coordinates. This is shown in Fig. 11. In addition a molar volume term was also incorporated as suggested by previous work (Jacques and Wyzgoski, 1979). The results shown in Fig. 11 are not particularly satisfying. For example many liquids in the cross-hatched region exhibit higher critical strains than anticipated based upon their solubility parameter differences with nylon.

Both Figures 10 and 11 indicate a common feature. That is, liquids which are absorbed to a large extent (Fig. 10) or which should be highly compatible with nylon (Fig. 11) can exhibit higher critical strains than expected. In a few

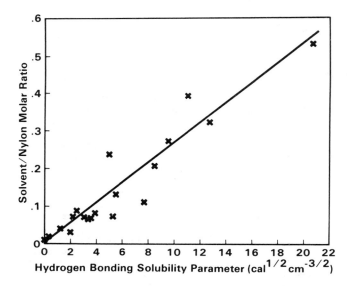

Fig. 9.   Ratio of moles of liquid absorbed per mole of nylon
          6,6 monomer versus hydrogen bonding solubility
          parameters of the absorbed liquid.

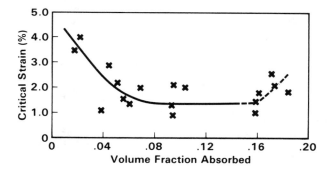

Fig. 10.   Critical strains versus corresponding volume
           fraction absorbed for nylon 6,6 in organic
           liquids.

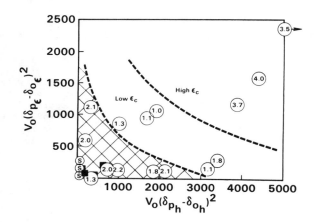

Fig. 11. Critical strain – Solubility parameter map for
nylon 6,6 in organic solvents. Numbers indicate
critical strains (percent) and alcohols are
represented by squares.

instances a similar phenomenon has been reported for crazing of glassy polymers by
Kambour and Gruner (1978). For example liquids which are highly absorbed can
reduce the glass transition temperature to below the test temperature and thereby
plasticize the surface. For glassy polymers such liquids can actually dissolve
the outer surface giving a tacky surface during the test. In such cases craze
initiation is suppressed due to the rapid homogeneous stress relaxation of the
surface. In nylon however the crystalline phase would be expected to prevent
complete dissolution of the swollen surface though stress relaxation would still
occur.

To determine whether surface plasticization was a plausible explanation for the
unexpectedly high critical strains for nylon a series of alcohols was investigated in more detail. This included methanol, ethanol, propanol, and butanol.
These liquids are represented by the partially hidden solid squares in Fig. 11.
Table 4 compares the critical strains for this series with the equilibrium weight
gains measured for thin films at 50°C. The question is why does methanol exhibit
a higher critical strain though also showing a larger solubility in the nylon. If
this is caused by a more rapid plasticization of nylon by methanol then it should
be possible to demonstrate that methanol absorption is more rapid and that stress
relaxation of nylon in methanol is more rapid.

TABLE 4   Critical Strains and Equilibrium Weight Gains for Nylon 6,6 in Alcohols

Liquid	ε_c (%)	Weight Gain (%)
Methanol	2.5	11.8
Ethanol	1.9	11.1
Propanol	1.1	11.1
Butanol	1.0	7.5

Thin films at 50°C immersed in alcohols were already equilibrated when the first
weight gain measurements were made. Therefore differences in the rate of absorption of the alcohols in nylon could not be defined from this data. Additional
data were obtained at 40°C using thicker films (0.15 cm) to decrease the rate of
solvent uptake. Fig. 12(a) shows how the rate of absorption decreases with
increasing alcohol chain length. The diffusion coefficients listed in Table 5
were measured by replotting the data on a square root of time basis (Fickian
behavior). As shown in Fig. 12(b) the data are reasonably fit with a straight
line except for an initial offset on the time axis. The offset suggests that the
sorption is nonFickian however it cannot be completely described as Case II.
Since relatively thick samples were employed it is likely to be a combination of
Fickian and Case II behavior (Thomas and Windle, 1978). For the purpose of the
present discussion the calculated "Fickian" diffusion coefficients serve to
demonstrate that absorption of methanol is much more rapid than absorption of the
higher molecular weight alcohols.

TABLE 5   Diffusion Coefficients for Alcohols in Nylon 6,6 at 40°C

Liquid	$D\ (x\ 10^{-9}\ \frac{cm^2}{sec})$
Methanol	35.7
Ethanol	1.8
Propanol	0.22
Butanol	0.07

Fig. 13 compares the stress relaxation behavior of nylon in air and in alcohols.
Since the samples are 0.15 cm thick only the outer surface will be rich in
methanol considering the relatively short measurement time. Still it is clear
that methanol immersion results in a more rapid decrease in stress and this is
attributed to the more rapid absorption of methanol. For propanol, on the other
hand, the relaxation rate is similar to that in air. Thus the stress relaxation
measurements confirm what is expected based upon the weight gain data. Namely,
stress relaxation is not accelerated in propanol because very little propanol is
absorbed during the time frame (20 minutes) of the measurement.

As an independent measure of the extent to which absorbed alcohol plasticizes
nylon 6,6 the decrease in the glass transition temperature was measured. Dynamic
mechanical measurements were employed for this purpose. Fig. 14 displays the
relative rigidity - temperature data for nylon 6,6 after equilibration in the
various alcohols. Glass transition temperatures were taken as the onset of the
transition from the glassy (high frequency) to rubbery (low frequency) state.
These are listed in Table 6. The low values observed demonstrate that nylon 6,6
which has been equilibrated in any of the alcohols is far above Tg at room temperature, the normal test temperature. No distinction can be made for the
crazing behavior of the alcohols based upon the relation of the test temperature
to the Tg of the equilibrated film. However these results do serve to point out
that methanol is much more effective in lowering the Tg of nylon 6,6 (even though
weight gains are similar) compared to ethanol and propanol. Thus both the more
rapid rate of absorption and the more plasticized nature of nylon 6,6 in methanol
can be contributing to the more rapid relaxation of surface stresses.

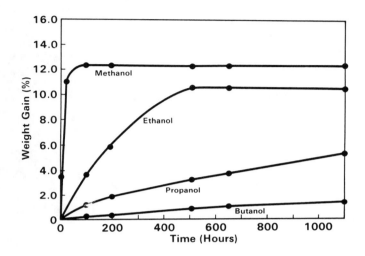

Fig. 12(a).   Weight gain versus time for nylon 6,6 at 40°C
in various alcohols.

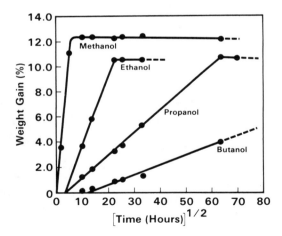

Fig. 12(b).   Sorption data for nylon 6,6 in alcohols
plotted on a square root of time basis.

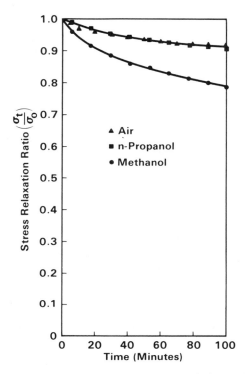

Fig. 13. Stress relaxation of nylon 6,6 in air compared
          to that in alcohols at room temperature.

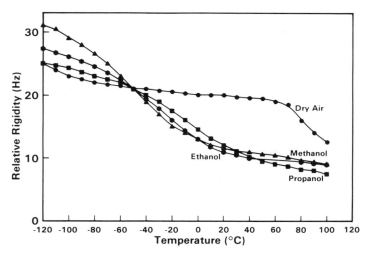

Fig. 14. Frequency (relative rigidity) versus temperature
          for nylon 6,6 films previously equilibrated in
          alcohols at 50°C.

TABLE 6   Glass Transition Temperature of Nylon 6,6 Films
Equilibrated in Alcohols at 50°C

Sample Conditioning	Tg (°C)
Dry Air	70
Methanol	−85
Ethanol	−67
Propanol	−44

To shed further light on the effect of alcohols on craze growth, measurements
were made of the craze propagation rate under constant load. In this case a razor
cut precrack was introduced to provide identical craze initiation conditions.
The experimental setup will be described in detail in a subsequent report. Fig.
15 shows a view of a large craze growing across the nylon sample immersed in the
test liquid. Since little or no crack growth occurs the time to failure is
determined almost entirely by the craze propagation rate. Fig. 16 shows time-to-
fail data for methanol and butanol at various initial stress intensity values.
The two liquids give identical results. This demonstrates clearly that the craze
propagation rate is similar for both liquids and therefore differences in critical
strain must reflect differences related to the initiation of crazes, not their
growth. In other words the surface plasticization by methanol suppresses craze
initiation by allowing a homogeneous stress relaxation at surface flaws. However,
apparently once a craze is initiated (by increasing the strain level), its growth
is not significantly altered by the surface plasticization.

A possible consequence of the competition between surface crazing and surface
plasticization is that factors which preferentially retard the kinetics of absorption (thereby slowing the plasticization of the surface) may increase the likeli-
hood of craze formation. To examine this hypothesis additional weight gain measurements were conducted at −40°C. The rate of solvent uptake was drastically
reduced as shown in Fig. 17. The corresponding critical strains measured at −40°C
are compared with room temperature measurements in Table 7. The lower critical
strain at −40°C for nylon 6,6 in methanol indicates that craze initiation is not
retarded by the surface plasticization process at this low temperature. It is
interesting to note that at −40°C the diffusion coefficient measured (again
assuming Fickian behavior) for methanol in nylon 6,6 is 0.006×10^{-9} cm^2/sec.
This value is lower than the diffusion coefficient calculated for butanol at 40°C.
Thus the similar behavior of methanol at −40°C and butanol at room temperature can
be understood. A decrease in critical strain with decreasing temperature is also
noted in comparing data for acetone and methyl isobutyl ketone. Acetone, like
methanol, is rapidly absorbed by nylon at room temperature whereas methyl isobutyl
ketone is slowly absorbed (see Fig. 5). Therefore the role of solubility in
stress cracking may be modified by the sorption kinetics over a wide range of
solubilities, not just for highly compatible liquids.

Fig. 15.   View of two large crazes growing from precracks
            in a nylon 6,6 sample under constant tensile load.

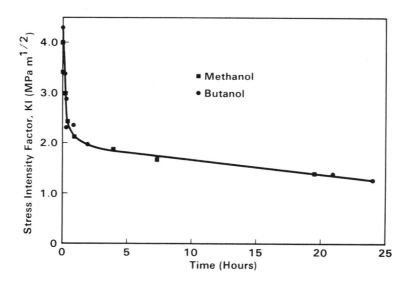

Fig. 16.   Initial stress intensity factor versus time-to-
            fail for nylon 6,6 immersed in methanol or
            butanol under constant load.

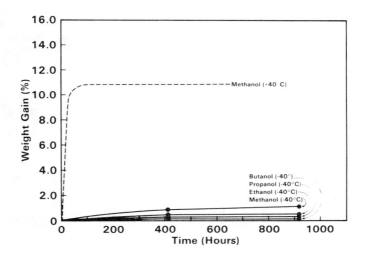

Fig. 17. Weight gain versus time for nylon 6,6 immersed in
          alcohols at −40°C.

TABLE 7   Effect of Temperature on Critical Strains of Dry Nylon 6,6

Liquid	Critical Strain	
	21°C	−40°C
Methanol	2.5	1.0
Ethanol	1.9	1.0
Propanol	1.1	1.0
Butanol	1.0	1.0
Acetone	2.2	1.0
Methyl isobutyl ketone	1.1	1.2

SUMMARY

The results of this study indicate that the role of solubility in stress cracking
of nylon 6,6 is not unlike that previously observed for glassy amorphous polymers.
Liquids which are highly soluble in nylon 6,6 are potentially more severe stress
cracking agents. The distinguishing feature of nylon 6,6 is its capacity to
rapidly absorb many organic solvents thereby allowing homogeneous stress relaxation to compete with craze and crack initiation. This in part may reflect the
relatively low glass transition temperature of dry nylon (Tg by DSC is 60°C), but
also may relate to the presence of hydrogen bonding. The beneficial effect of

absorbed moisture may also be due to the rapid relaxation of stresses in the rubbery (Tg below room temperature)' surface of moisturized nylon 6,6. Solubility parameters can be used to estimate the interaction expected between nylon and organic liquids. However, in general both the solubility and sorption kinetics must be known in order to predict the stress cracking behavior of nylon 6,6 in organic liquids.

REFERENCES

Askadskii, A. A., L. K. Kolmakova, A. A. Tager, G. L. Slonimskii, and V. V. Korshak (1977). The Assessment of the Cohesive Energy Density Between Low Molecular Weight Liquids and Polymers. Polym. Sci: USSR, A19: No. 5, 1004-1013.
Barton, A. F. M. (1975). Solubility Parameters. Chem. Reviews, 75, 731-753.
Burrell, H. (1955). Solubility Parameters. Interchemical Reviews, 14(3), 3-46.
Dunn, P. and G. F. Sansom (1969a). The Stress Cracking of Polyamides by Metal Salts. Part I. Metal Halides. J. Appl. Polym. Sci., 13, 1641-1655.
Dunn, P. and G. F. Sansom (1969b). The Stress Cracking of Polyamides by Metal Salts. Part II. Mechanism of Cracking. J. Appl. Polym. Sci., 13, 1657-1672.
Dunn, P. and G. F. Sansom (1969c). The Stress Cracking of Polyamides by Metal Salts. Part III. Metal Thiocyanates. J. Appl. Polym. Sci., 13, 1673-1688.
Dunn P. and G. F. Sansom (1970). The Stress Cracking of Polyamides by Metal Salts. Part IV. Metal Nitrates. J. Appl. Polym. Sci., 14, 1799-1806.
E. I. du Pont de Nemours and Co. (Inc.) (1980). Zytel Nylon Resin. Design Handbook, 66.
Jacques, C. H. M. and M. G. Wyzgoski (1979). Prediction of Environmental Stress Cracking of Polycarbonate from Solubility Considerations. J. Appl. Polym. Sci., 23, 1153-1166.
Kambour, R. P. (1973). Crazing of Plastics. J. Polym. Sci., Part D, Macrom. Reviews, 7, 1-154.
Kambour, R. P. and C. L. Gruner (1978). Effects of Polar Group Incorporation on Crazing of Glassy Polymers: Styrene-Acrylonitrile Copolymer and a Dicyano Bisphenol Polycarbonate. J. Polym. Sci. (Physics), 16, 703-716.
Kramer, E. J. (1979). Environmental Cracking of Polymers. In E. H. Andrews (ed), Developments in Polymer Fracture, Applied Science, London, 55-120.
Rabinowitz, S. and P. Beardmore (1972). Craze Formation and Fracture in Glassy Polymers. CRC Critical Reviews in Macromolecular Sciences, 1, 1-45.
Rogers, C. E., V. Stannett, and M. Szwarc (1959). The Sorption of Organic Vapors by Polyethylene. J. Phys. Chem., 63, 1406-1413.
Thomas, N. and A. H. Windle (1978). Transport of Methanol in Poly(methyl methacrylate). Polymer, 19, 255-265.
Weiske, C. D. (1964). Chemical Resistance and Stress Cracking of Polyamides (German). Kuntstoffe, 54, 626-634.
Wyzgoski, M. G. and C. H. M. Jacques (1977). Stress Cracking of Plastics by Gasoline and Gasoline Components. Polym. Engr. and Sci., 17, 854-860.

ACKNOWLEDGMENT

Discussions of the solvent uptake data with Prof. Harold Hopfenberg of North Carolina State University were most helpful.

SOLUTION BEHAVIOR OF GROUP IV B POLYOXIMES

Charles E. Carraher, Jr. and Larry P. Torre
Wright State University
Department of Chemistry
Dayton, OH   45435

ABSTRACT

The effect of polymer chain length, copolymer content, metal and symmetry of oxime on polymer solubility are studied for Group IV B polyoximes using chain termina-tors, model compounds and copolymers. Polymer solubility is largely independent of chain length but is dependent on the nature of the metal and symmetry of the dioxime. The general poor solubility appears inherent in the repeat unit struc-ture.

KEYWORDS

Organometallic monomers, organometallic polymers, dioxime, oxime, dicyclopenta-dienylhafnium dichloride, polymer solubility, cocondensation polymers, Vitamin K_3, 2-phenyl-1,3-indanone, 1,4-cyclohexanedione dioxime, steroids, steroid di-oximes

INTRODUCTION

The poor solubility of condensation organometallic polymers has been a constant problem and an area of concern (1,2). Recently we reported the synthesis of Group IV B polyoximes (3). A major purpose of this study involved a systematic study of selected parameters with regard to polymer solubility.

$$Cp_2MCl_2 + R\overset{\displaystyle HON}{\underset{\|}{C}}-R'-\overset{\displaystyle NOH}{\underset{\|}{C}}-R \longrightarrow \left(O-N=\overset{\displaystyle R}{\underset{|}{C}}-R'-\overset{\displaystyle R}{\underset{|}{C}}=N-O-\overset{\displaystyle Cp}{\underset{|}{M}}\right) \tag{1}$$

The general poor solubility of organometallic polymers is believed to be due to a combination of factors including chain stiffness, crystallization tendencies and the presence of a unique combination of highly polar and nonpolar regions within a single unit. A number of ploys have been utilized in attempts to increase solubility. Here is considered the factors of chain length, product symmetry and unit structure and their effect on polyoxime solubility.

61

EXPERIMENTAL

Monomer synthesis

A solution of two equivalents of hydroxylamine, one equivalent of diketone, a
slight excess of two equivalents of base (pyridine) in absolute ethanol was
refluxed for three or more hours and evaporated to dryness (Sorenson, Campbell,
1968). The resulting solids were washed with 1:1 ethanol-water, and the product
was removed by suction filtration. The product was recrystallized twice from 1:1
ethanol-water.

2-Methyl-1,4-naphthaquinone dioxime (menadioxime, Vitamin K_3 dioxime), (Rappoport, 1967): the yellow solid was obtained in 49% yield; m.p. = 164-167°C, (lit.
166-168°C); IR (KBr) 3260 cm^{-1} (O-H), 1620 cm^{-1} (C=N), 980 cm^{-1} (N-O).

Δ^1-Pregnan-3,20-dioxime (progesterone dioxime), (Rappoport, 1967): the white
solid was obtained in 86% yield; m.p. = 248-252°C, (lit. 247-256°C); IR (KBr) 3280
cm^{-1} (O-H), 1640 cm^{-1} (C=N), 915-980 cm^{-1} (N-O).

$\Delta^{1,4}$-Androstadien-3,17-dioxime (androstadiendioxime), (Hershberg, 1948) the
white solid was obtained in 90% yield; m.p. = 232-235°C, (lit. 229-230°C); IR
(KBr) 3260 cm^{-1} (O-H), 1660 cm^{-1} (C=N), 870-980 cm^{-1} (N-O).

3,20-Pregnanedioxime (pregnanedioxime), (Jarrot, Laine, Qui, Goutarel, 1962):
the white solid was obtained in 72% yield; m.p. = 174-176°C, (lit. 260°C); IR
(KBr) 3300 cm^{-1} (O-H), 1695 cm^{-1} (C=N), 900-1010 cm^{-1} (N-O),

Anal. Calcd. for $C_{21}H_{34}O_2N_2$: C, 73.0; H, 9.9; O, 9.3%

Found: C, 76.2; H, 10.1; O, 9.2%

1,4-Cyclohexanedione dioxime, (Pollock, Stevens, 1965): the white crystalline
solid was obtained in 70% yield; m.p. = 189-192°C, (lit. 188°C), IR (KBr) 3200cm^{-1}
(O-H), 1660 cm^{-1} (C=N), 925-1000 cm^{-1} (N-O).

2-Phenyl-1,3-indanone dioxime (phenindioxime), (Pollock, Stevens, 1965): the white
solid was obtained in 68% yield; m.p. = 198-200°C, (lit. 193-196°C; IR (KBr) 3200
cm^{-1} (O-H), 1650 cm^{-1} (C=N), 957, 1045 cm^{-1} (N-O).

Reaction procedure

The reaction vessel for all polymerization reactions was a one pint Kimex Emulsi-
fying jar fitted onto a Waring Blendor 700 (Model 31BL46).

Interfacial syntheses were performed by adding an aqueous solution of dioxime,
typically 1.0 mmole of dioxime dissolved in 25.0 ml of water containing 2.0 mmoles
of sodium hydroxide, to the reaction vessel. To overcome a surface wetting
problem, the dioxime was dissolved into solution by rapid stirring in the reaction
vessel. The stirring was stopped after 30 seconds leaving a clear solution.
Stirring was begun and an organic solution containing the Group IVB metallocene
dichloride, typically 1.0 mmole of metallocene dichloride per 25.0 ml of chloro-

form, was added through the powder funnel. Timing was begun after completion of addition of the organic phase. Total time of addition was three seconds or less. Usually stirring was stopped after 60 seconds, and the reaction mixture, typically a white to yellow gel, was filtered using a Buchner filter with suction. The filtered product was washed with water, chloroform, and water again (50 ml of each) to remove unreacted monomers. The gelatinous product was then washed with water into a pre-weighed glass petri-dish, and left to dry in air at room temperature.

Aqueous solution syntheses were performed by adding an aqueous solution of dioxime (1.0 mmole dioxime in 25.0 ml water) to a rapidly stirred (23,000rpm, no-load) aqueous solution of metallocene dichloride (1.0 mmole metallocene in 25.0 ml water). Product isolation and treatment are described directly above.

Model compounds of the Group IVB metal containing polyoximes were prepared by adding an aqueous solution of cyclohexanone oxime (2.0 mmole cyclohexanone oxime in 25.0 ml water containing 2.0 mmole of NaOH), to a rapidly stirred solution of metallocene (1.0 mmole metallocene in 25.0 ml of chloroform). Product isolation and treatment are as described above.

Polymer Solubility

The polyoximes of the present study were tested by placing about 2mg of polyoxime in 2-3 ml of test solvent with shaking. The liquid-polymer mixtures were observed over a period of a week. Disappearance of the polymer or coloration of the test liquid were taken as indication that dissolution had occurred. The test solvents used included organic solvents (benzene, chloroform, hexane, carbon tetrachloride, etc.), water, and the following dipolar aprotic solvents: dimethylsulfoxide, dimethylformamide, dimethylacetamide, hexamethylphosphoramide, triethyl phosphate, and 1-methylimidazole. The polyoximes were slightly soluble in dimethylacetamide and 1-methylimidazole, with the latter being the best solvent. All solubility testing and molecular weight determinations were made on polyoximes in 1-methylimidazole. Solubility of some of the polyoximes is given in Table 1 in units of % polyoxime (by weight) and grams polyoxime per 100 ml 1-methylimidazole. Solubility was determined by adding about 50 mg of polyoxime to 40 ml of 1-methylimidazole with stirring. After six hours the solution was filtered through a sintered glass funnel three times to remove any (typically 10%) undissolved material. If the polyoxime was dissolved, additional polymer was added and the sequence repeated. Additional solvent was added to undissolved material and the procedure repeated.

Physical characterization

Infrared (IR) spectra were obtained using potassium bromide pellets with a Perkin-Elmer Model 457 Grating Spectrophotometer.

Light scattering photometry was conducted utilizing serial dilutions employing a Brice-Phoenix BP-3000 Universal Light Scattering Photometer. Refractive Index Increments were determined using a Bausch and Lomb Abbe Refractometer Model #3-L.

Elemental analyses were performed by Galbraith Laboratories, Knoxville, Tenn. Elemental analyses for Group IV B metals were performed by thermal degradation of the samples to metal oxides, MO_2.

RESULTS AND DISCUSSION

General Solubility

The solubility of organometallic polymers has been a constant problem (Carraher, Sheats, Pittman, 1981; Carraher, 1977). Related to the present study, previously synthesized Group IVB metal containing polyesters, polyethers, polyamidoximes, and polyferroceneoximes exhibited low degrees of solubility or were insoluble (Carraher, Sheats, Pittman, 1981; Carraher, Bajah, 1975; Carraher, Christensen, 1978; Carraher, Frary, 1974).

The polyoximes are insoluble in all attempted liquids except dimethylacetamide and 1-methylimidazole, the latter being the best solvent with respect to both breadth and extent. Table 1 contains solubility results for a number of polyoximes in 1-methylimidazole. Solubility is dependent on the nature of the metal and oxime portion but no regular trend is apparent and solubility is low for all of the polyoximes.

Table 1 Polyoxime Solubility in 1-Methylimidazole

| Polyoxime | Solubility | |
	% (by weight)	g/100 ml solvent
Δ^1-Pregnan-3,20-dioxime + Cp_2Ti	0.60	0.62
Δ^1-Pregnan-3,20-dioxime + Cp_2Zr	0.43	0.44
Δ^1-Pregnan-3,20-dioxime + Cp_2Hf	0.10	0.11
3,20-Pregnanedioxime + Cp_2Ti	0.067	0.069
3,20-Pregnanedioxime + Cp_2Zr	0.042	0.043
$\Delta^{1,4}$-Androstadiendioxime + Cp_2Ti	0.13	0.15
$\Delta^{1,4}$-Androstadiendioxime + Cp_2Zr	0.18	0.18
2-Methyl-1,4-naphthoquinone dioxime + Cp_2Ti	0.27	0.28
2-Methyl-1,4-naphthoquinone dioxime + Cp_2Zr	0.079	0.081
2-Methyl-1,4-naphthoquinone dioxime + Cp_2Hf	0.033	0.034
2-Phenyl-1,3-indanone dioxime + Cp_2Ti	0.21	0.23
2-Phenyl-1,3-indanone dioxime + Cp_2Zr	0.39	0.40
p-Benzoquinone dioxime + Cp_2Ti	0.11	0.11
p-Benzoquinone dioxime + Cp_2Zr	0.026	0.033
1,4-Cyclohexane dioxime + Cp_2Ti	0.2	0.22

Model Compounds

In an attempt to determine the predominate factor(s) governing poor solubility (such as polymer chain size or polymer structure), model compounds including the form 2 were synthesized. Past work with model compounds of Group IVB metallocene containing polymers have shown poor solubility to be an inherent property of compound structure (Carraher, Sheats, Pittman, 1981; Carraher, 1977; Carraher, 1972)).

$$2 \quad \underset{\text{NOH}}{\bigcirc} + Cp_2MCl_2 \xrightarrow{\text{base}} \bigcirc =N-O-\underset{\overset{|}{Cp}}{\overset{Cp}{M}}-O-N= \bigcirc \tag{2}$$

M = Ti, Zr, Hf

Solubility of the model compounds in 1-methylimidazole and the other tested liquids were of a low degree, essentially the same as for the corresponding poly-oxime (Table 2). Thus it is consistent with the poor polyoxime solubility being, at least in part, due to factors inherent in the polymer structure and not necessarily polymer chain length. What specific factors are responsible is unknown and more study must be made to clarify this matter.

In some measure the above observations must remain uncertain since the repeat unit in the polymer is a A-B-A-B-A-B type (where A = oxime moiety, B = metallocene moiety) whereas the repeat unit for the model compound is A-B-A.

Dissymmetrical Monomers

A primary consideration in the selection of most of the diketones (and thus the corresponding dioximes) was their large size and the geometrical dissymmetry of the molecule. Figure 1 contains structures of the utilized dioximes. The steroid dioximes are relatively large. The Vitamin K_3 and 2-phenyl-1,3-indanone dioximes are of a more moderate size compared to the steroids. p-Benzoquinone dioxime and 1,4-cyclohexanedione dioxime are, in comparison, small. All but the 2-phenyl-1,3-indanone, p-benzoquinone and 1,4-cyclohexanedione dioximes are disymmetrical giving rise to further dissymmetry in the actual polymer chain.

Table 2   Model Compound Solubilities in 1-Methylimidazole

Model Compound	Solubility (% by wt.)
$\bigcirc =NO - \underset{\overset{|}{Cp}}{\overset{Cp}{Ti}} - ON= \bigcirc$	0.38
$\bigcirc =NO - \underset{\overset{|}{Cp}}{\overset{Cp}{Zr}} - ON= \bigcirc$	0.52
$\bigcirc =NO - \underset{\overset{|}{Cp}}{\overset{Cp}{Hf}} - ON= \bigcirc$	0.60

If either or both factors of Lewis base size and lack of symmetry are important factors in determining polyoxime solubility, then the polyoximes containing the

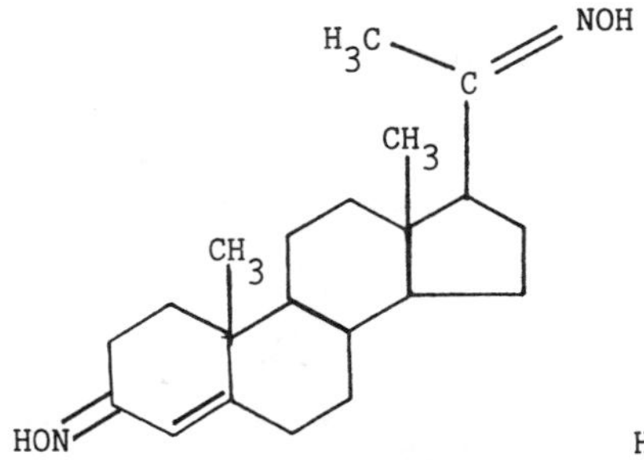

Fig. 1. Structures of Dioximes Employed in the Synthesis.

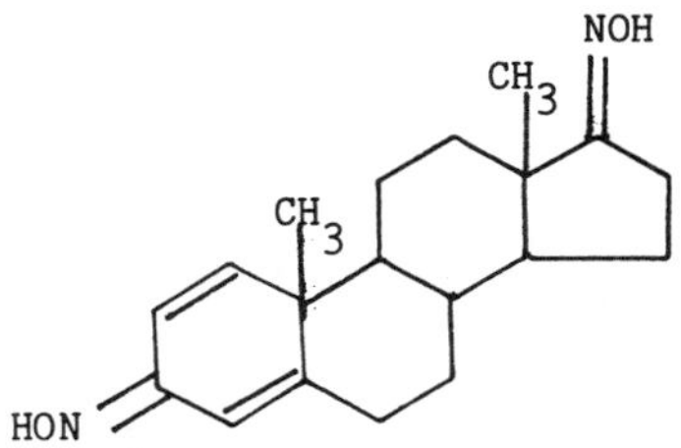

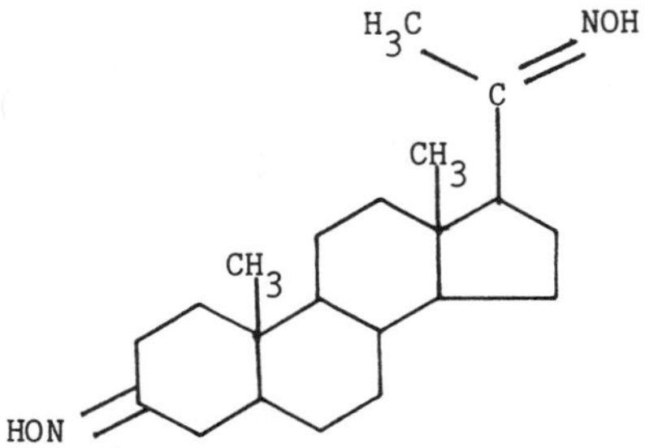

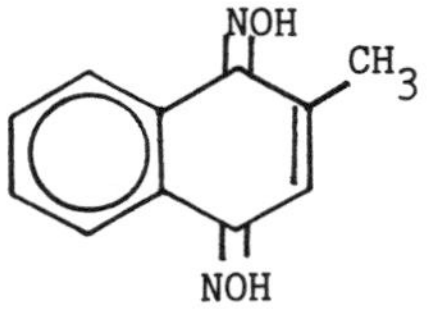

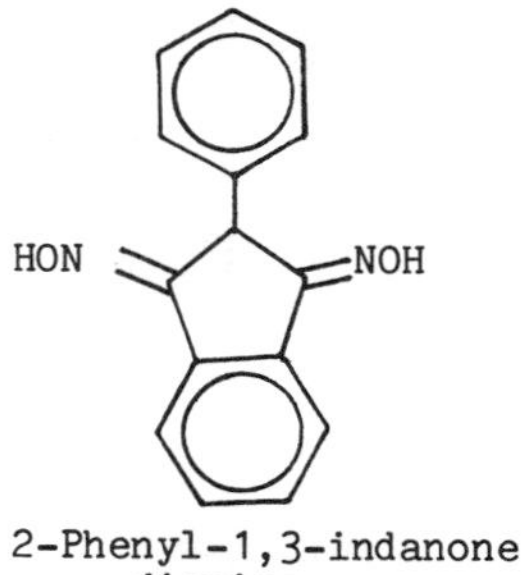

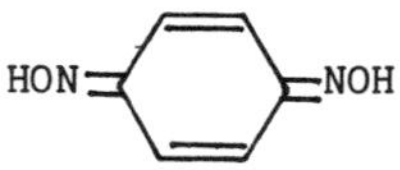

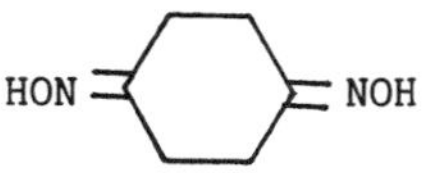

steroids should be more soluble than the polyoximes containing the Vitamin K_3 or indanone moieties, which should be more soluble than the symmetric, linear p-benzoquinone and 1,4-cyclohexanedione dioxime moiety containing polyoximes. This is generally consistent with the observed trend. It is not to be inferred that the factors of size and/or lack of molecular symmetry are the only factor(s) in determining polymer solubility. Again, more study must be made of this subject before factors affecting organometallic polymer solubility can be identified and interrelated.

Molecular Weight

A major factor often affecting polymer solubility is the polymer chain length, with the trend generally being the larger the chain, the lower the solubility of that chain.

Table 3 contains results of solubility as a function of molecular weight. There does not appear to be a correlation between molecular weight and solubility.

Table 3 Polyoxime Solubility in 1-Methylimidazole as a Function of Chain Length

Polyoxime	dn/dc	MW (g/mole)	DP (units)	Solubility (g/100 ml solvent)
Δ^1-Pregnan-3,20-dioxime + Cp_2Ti	-0.217	3.0×10^5	592	0.62
Δ^1-Pregnan-3,20-dioxime + Cp_2Zr	-0.631	1.61×10^4	29	0.44
3,20-Pregnanedioxime + Cp_2Ti	-2.985	9.68×10^3	19	0.069
$\Delta^{1,4}$-Androstandiendioxime + Cp_2Ti	-1.923	1.99×10^4	41	0.15
2-Methyl-1,4-naphthoquinone dioxime + Cp_2Ti	-1.053	9.08×10^4	240	0.28
2-Phenyl-1,3-indanone dioxime + Cp_2Ti	-0.664	1.58×10^5	369	0.23
p-Benzoquinone dioxime + Cp_2Ti	-1.944	2.10×10^4	67	0.11
1,4-Cyclohexanedioxime + Cp_2Ti	-1.244	7.82×10^3	25	0.22

In an attempt to shorten the polyoxime chain length (and thereby increase the polymer solubility), a reactant capable of acting as a chain terminator was introduced (in varying amounts) into the polymerization reaction. The chain terminator was cyclohexanone oxime which was used because it was readily available and is similar in structure to the 1,4-cyclohexanedione dioxime monomer which had already been established to give product in reasonable yield. However, Δ^1-pregnan-3,20-dioxime + Cp_2Ti polyoximes synthesized in the presence of from 1 to 40% (mole %) cyclohexanone oxime yielded products which showed no appreciably different solubility properties than the same polymer synthesized without added chain terminator. It may be that the low reactivity of cyclohexanone oxime prevented the inclusion of that moiety in the rapid polycondensation. (Cyclohexanone oxime can be seen to have a low relative reactivity as the model compounds (2) had product yields of 10 to 20%.) Even so there was enough chain terminating agent present to (supposedly) effectively shorten the polyoxime chains. This is consistent with the poor solubility of the model compounds noted

in a previous section.   Again, more study of this approach to increase polymer
solubility must be made.

Copolymers

Borden (1978) has shown that the use of copolymerization reactions employing
organic acid chlorides along with Group IVB containing Cp_2MCl_2 in reaction with
diamines and diols apparently lowers chain symmetry giving copolymers which are
soluble in dipolar aprotic solvents as well as chloroform.

Copolymers A and B (Table 4) were synthesized as described.   The ferrocene containing dioxime was utilized because it gives a ready method for determining the
copolymer composition by analysis for iron.   Elemental analysis for iron gave
results consistent with a 1:1 incorporation of each dioxime in the polyoxime
chain.   (Copolymer A, assuming a 1:1 inclusion of both dioximes, would give a %Fe
of 11.8; %Fe found was 11.2%.) Light scattering analysis of the copolyoximes gave
weight average molecular weights within the range found for the homopolyoximes.
Full characterization of these copolymers was not conducted since it was only
their solubility properties that were of primary interest.

Table 4   Copolymer Solubilities

	Solubility (% by weight)
Δ^1-Pregnan-3,20-dioxime + Cp_2Ti	0.60
Diacetylferrocene dioxime + Cp_2Ti	0.14
Copolymer A	0.88
Δ^1-Pregnan-3,20-dioxime + Cp_2Ti	0.60
Dibenzoylferrocene dioxime + Cp_2Ti	0.039
Copolymer B	0.40

Copolymer A had a solubility in 1-methylimidazole of 0.88% by weight.   The solubilities of the analogous homopolymer from Δ^1-pregnan-3,20-dioxime and the analogous homopolymer from the diacetylferrocene dioxime (Table 3) were both lower
showing that the synthesis of the copolymer did increase solubility.

However, copolymer B had a solubility in 1-methylimidazole of 0.40% by weight.   As
seen in Table 3, this copolymer was less soluble than the associated homopolyoximes.   Thus it appears that each copolymer-homopolymer combination must be
studied separately with regard to solubility.

SUMMARY

A factor which appears to affect polyoxime solubility is the symmetry of the
utilized oxime such that the use of dissymmetrical oximes enhances product solubility.   Solubility appears to be (largely) independent of chain length.   The
general poor solubility appears to be inherent in the repeat unit structure.

REFERENCES

Borden, D.G. (1978). Photocrosslinkable Organometallic Polyester Polymers. In. C.
    Carraher, J. Sheats, and C. Pittman (Eds.),Organometallic Polymes, Academic
    Press, New York, pp. 115-128.

Carraher, C. (1972). Synthesis of Group IV Polymers by the Interfacial Technique. Inorg. Macromol. Revs., 1, 271-286.

Carraher, C. (1977). Organometallic Polymers, F. Millich and C. Carraher (Eds.), Interfacial Synthesis Vol II, Marcel Dekker, New York. pp. 367-416.

Carraher, C. and S. Bajah (1975). Effect of Base Nature, Base Concentration and Method of Synthesis of Titanium Polyethers. British Polymer J., 7, (155-159).

Carraher, C. and M. Christensen (1978). Interfacial Synthesis of Condensates of Dicyclopentadienyltitanium Dichloride and 1,1'-Diacylferrocene Oximes. Angew. Makromol. Chemie, 69, (61-66).

Carraher, C. and R. Frary (1974). Synthesis of Zirconium Poly(-0-amidoximes). J. Polymer Sci., 12(A1), (799-805).

Carraher, C., J. Sheats and C. Pittman (Eds.). (1981). Advances in Organometallic Polymers, Marcel Dekker, New York.

Carraher, C. and L.P. Torre (1980). Synthesis of Titanium Polyoxines Using the Interfacial Technique, Organic Coatings and Plastics Chemistry, 42, (18-22).

Hershberg, E.B. (1948). Regeneration of Steroid Ketones from Their Semicarbazones with Pyruric Acid, J. Org. Chem., 13, (542-546).

Janot, M., F. Laine, K-H. Qui and R. Goutarel (1962). Steroid Alkaloids. IX. 20-Amino and 3,20-Diamino Derivatives of 5-Alpha-Pregnane, Bull. Soc. Chim. France, 111-118.

Pollock, J.R. and R. Stevens (Eds.) (1965). Dictionary of Organic Compounds, Vol. IX, 4th ed. Oxford University Press, New York. pp. 785 and 2699.

Rappoport, Z. (Compiler) (1967). Handbook of Tables for Organic Compound Identification. CRC, Cleveland, Ohio. pp. 181 and 2779.

Sorenson, W. and T. Campbell (1968). Preparative Methods of Polymer Chemistry, 2nd ed. Interscience, New York. p. 118.

ALIPHATIC STEARATES OF INCREASING CHAIN LENGTH :
CHARACTERIZATION AND SOLUBILITY IN POLYVINYLCHLORIDE

M. Pizzoli[*], G. Pezzin[*], G. Ceccorulli[*], M. Scandola[*] and
G. Crose[**]

[*]Centro di Studio di Fisica Macromolecolare del C.N.R.,
Istituto "Ciamician", Via Selmi 2, Bologna, Italy
[**]Montepolimeri, Porto Marghera, Venezia, Italy

ABSTRACT

The Methyl-, Ethyl-, n-Butyl-, n-Octyl- and n-Hexadecyl esters of Octadecanoic
(Stearic) acid were synthetized from the corresponding alcohols and pure Stearic
acid. Their physical properties were determined and their viscous flow parameters
were calculated according to both the Eyring theory and the Batschinski-Hildebrand
treatment of viscosity. The results are favourably compared with similar published
data on hydrocarbons and with values for low molecular weight esters. Different
amounts of each stearate were mixed with Polyvinylchloride at 90°C. The solubility
of the esters in the polymer, determined by measuring the crystallizable undissolved
ester by means of a Differential Scanning Calorimeter, decreases from 6.5% to 0.3%
with increasing chain length of the stearates. The Flory-Huggins interaction
parameter χ, obtained from the solubility results, increases from 1.6 (Methylstearate)
to 4.2 (n-Hexadecylstearate). The lengthening of the aliphatic chain, which results
in a dilution of the ester groups, apparently weakens the PVC-ester interactions.
The glass transition temperature T_g decreases with increasing diluent content down
to a constant value closely related to the ester solubility. The decrease of T_g
brought about by the PVC-ester interactions is described by the free volume theory,
using a single equation for all the systems examined.

KEYWORDS

Polyvinylchloride; stearates; viscosity parameters; solubility; glass transition;
thermodynamic interaction parameter.

INTRODUCTION

Some of the data concerning the solubility of low molecular weight substances in
Polyvinylchloride (PVC) were obtained many years ago. Doty and Zable (1946)
calculated the Flory-Huggins interaction parameter χ for a number of solvents from
swelling measurements on cross-linked PVC and Anagnostopoulos and Koran (1962) used
the melting point depression method to evaluate χ for many plasticizers, although
the melting temperature of PVC is variously quoted as lying between 174°C (Anagnostopoulos and Koran, 1962) and 260°C (Ceccorulli, Pizzoli and Pezzin, 1977). A lack
of data on the high temperature solubility of PVC additives such as plasticizers,
stabilizers, lubricants, processing aids etc., is however apparent, so that a

70

systematic investigation on the system PVC-aliphatic esters was recently undertaken in our laboratory (Ceccorulli and co-workers, 1981).

In this paper, we report the results obtained in an investigation on the system PVC-stearates in which linear aliphatic esters of different molecular weight were synthetized, characterized and added to PVC. The solubility in the polymer at high temperature, as well as the glass transition temperature of the solvated system, were determined and found to be related to the different chain length of the alcoholic residue of the additive. To determine the solubility of additives in PVC we used a simple and reliable method based on the calorimetric determination of the (crystallizable) "excess" additive, a method widely used for hydrated macromolecular systems, such as proteins (Ceccorulli, Scandola and Pezzin, 1977).

In order to test the applicability of the current theories of viscosity, the viscometric and volumetric data of the pure stearates have been treated in terms of both the Eyring and Hildebrand equations (Glasstone, Laidler and Eyring, 1941; Hildebrand and Lamoreaux, 1972). Moreover, the decrease of the glass transition temperature brought about by the PVC-ester interactions has been satisfactorily described by a simple free volume theory (Braun and Kovacs, 1965).

EXPERIMENTAL

Synthesis and Purification of the Esters

For the preparation of the esters, a relatively pure Stearic acid was used. The content of Palmitic and Lauric acid was 0.7% and 0.8% respectively, and the melting point was 71°C. The alcohols used (Methyl, Ethyl, n-Butyl, n-Octyl and n-Hexadecyl) were obtained from two sources (C. Erba and Fluka) and all had 99% purity. The esterification was carried out using equimolecular mixtures of Stearic acid and alcohols, in the presence of p-Toluensulphonic acid (about 0.04 mol/l). The purity of the esters, controlled by gas-chromatography, was always better than 97%. The stearates were characterized by measuring their density and viscosity in a wide range of temperatures (from 40 to 160°C). Standard pycnometers and glass capillary viscometers were used. The data are collected in Table 1, together with the melting point and heat of fusion, determined with the Differential Scanning Calorimeter DSC2 at a scanning rate of 10 deg/min.

TABLE 1   Physical Properties of Stearates[*]

Substance	Melting Point[&] (°C)	Heat of Fusion (KJ/mol)	Density at 100°C (g/ml)	$\alpha = -\dfrac{d\ln\rho}{dT}$[∓] (°C^{-1})	Viscosity at 100°C (mPa·s)
Stearic acid	71	56.3	0.826	$0.86 \cdot 10^{-3}$	
Methylstearate	30	58.7	0.807	$0.89 \cdot 10^{-3}$	1.66
Ethylstearate	32	56.9	0.802	$0.90 \cdot 10^{-3}$	1.72
n-Butylstearate	26	50.4	0.802	$0.89 \cdot 10^{-3}$	1.98
n- Octylstearate	30	76.3	0.802	$0.84 \cdot 10^{-3}$	2.65
n-Hexadecylstearate	60	106.5	0.806	$0.96 \cdot 10^{-3}$	4.42

[*]From 40 to 160°C, with the exception of Stearic acid and n-Hexadecylstearate (from 70 to 160°C).
[&]Temperature of the melting endotherm peak.
[∓]Correlation coefficient: better than 0.9998.

Sample Preparation

The Polyvinylchloride used in this work was obtained by suspension polymerization
at 60°C. The intrinsic viscosity, in Cyclohexanone at 25°C, was 88 ml/g and the
chlorine content 56.2%. It was mixed with suitable amounts of a given stearate
(from zero to 0.3 diluent to PVC weight ratio) in a Brabender mixer, at 90°C and
60 rpm. The polymer was kept in the mixer for 5 min before addition of the stearate,
and the total mixing time was 35 min. The dry blended samples so obtained were
usually studied as such in the calorimeter, but in some cases they were calendered
at 150°C and 14 rpm for 5 min.

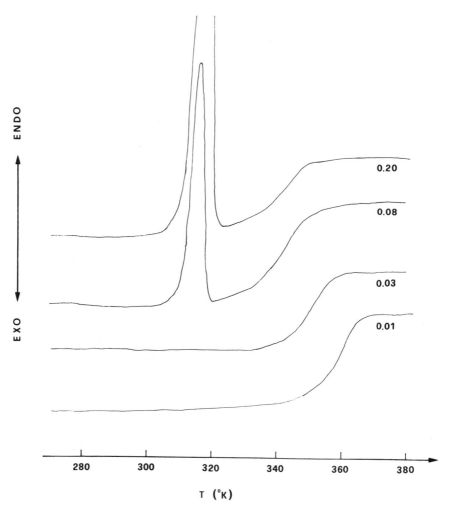

Fig. 1.   DSC curves for the system PVC-Methylstearate, at increasing
          stearate content (from 0.01 to 0.20 g/gPVC).

Calorimetric Measurements

A Differential Scanning Calorimeter DSC2 was used to monitore both the melting of the excess additive and the glass transition temperature T_g of the PVC-additive mixture. DSC curves were recorded from 50°C below the melting point of the pure additive to above the glass transition of the system, at a heating rate of 20°C/min. As previously reported (Ceccorulli and co-workers, 1981), at low additive concentrations, only the glass transition was revealed and its temperature was determined by the method of the baseline deviation. At higher concentrations, a melting endotherm due to additive present as a separate phase appeared. A typical example is shown in Fig. 1. In subsequent runs a slight decrease of the area under the melting curve was found for all systems, so that the samples were kept at 90°C for 120 min in the calorimeter in order to reach a constant value of the melting area. For most additives examined the melting endotherm is quite complex and shows a prominent peak preceeded by one or two shoulders. Multiple peaks are a rather common feature in the melting endotherm of both low (Ceccorulli, Scandola and Pezzin, 1977) and high (Ceccorulli, Manescalchi and Pizzoli, 1975) molecular weight substances. Although the shape of the endotherm was slightly affected by the crystallization conditions, the total area under the melting curve remained constant and the corresponding heat of fusion could be évaluated by calibration with a high purity standard. Due to the proximity of the melting and glass transition phenomena, at high additive content no reliable T_g data could be obtained for the system PVC-Stearic acid, also investigated in this work.

RESULTS AND DISCUSSION

Physical Properties of Stearates

As shown in Table 1, the melting temperature of the esters changes with molecular structure in a relatively narrow range, from 26 to 60°C, and the corresponding heat of fusion changes from 50.4 to 106.5 KJ/mol. The density at the fixed temperature of 100°C is substantially independent on molecular structure, the range of variation being 0.6%. A higher although non systematic variation of the expansion coefficient $\alpha = -d\ln\rho/dT$ is found; i.e., α shows a 14% increase going from n-Octyl stearate to n-Hexadecylstearate. The values reported, of the order of 10^{-3}°C^{-1}, are typical for the expansion coefficients of many organic liquids. The viscosity of the esters increases with increasing molecular weight, as shown in Fig. 2; a similar behaviour is found for the temperature dependence of viscosity, which follows an Arrhenius-type equation:

$$\eta = A \exp (E_a/RT) \tag{1}$$

E_a values, collected in Table 2, increase from 16.4 to 20.8 KJ/mol with increasing chain length of the esters.

The temperature dependence of the viscosity of liquids can be described with different equations, either empirical or derived from simplified models of the liquid state (Bondi, 1956). For example, the Arrhenius equation can be interpreted by the Eyring theory of viscosity (Ewell and Eyring, 1937), a model theory firstly proposed by Frenkel (1926). According to Eyring, the elementary flow event is the motion of a molecule (or segment) past its neighbours and into an adjacent "hole", a jump which is considered a passage over a potential barrier and therefore treated by the theory of absolute reaction rates. The flow rate is determined by the availability of the energy necessary to create locally enough free space for the translational jump, and the flow activation energy:

$$E_a = R \, d\ln \eta/d(1/T) \tag{2}$$

is interpreted accordingly as a measure of the "hole formation" energy. The fact
that for many liquids E_a is proportional to the enthalpy of vaporization is usually
taken as a strong argument in favour of the Eyring model (Glasstone, Laidler and
Eyring, 1941).

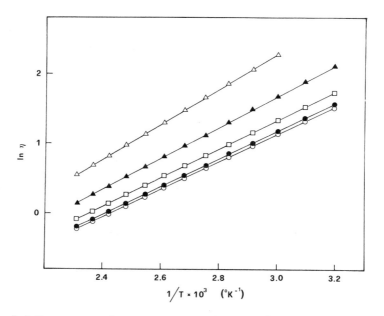

Fig. 2.   Temperature dependence of viscosity: $\bigcirc$ Methyl-; $\bullet$ Ethyl-;
          $\square$ n-Butyl-; $\blacktriangle$ n-Octyl- and $\triangle$ n-Hexadecyl-stearate.

TABLE 2   Viscous Flow Parameters of Stearates[*]

Substance	$E_a = R \dfrac{d\ln \eta}{d(1/T)}$ (KJ/mol)	$A \cdot 10^{-6}$ (Pa·s)	B (mPa·s)$^{-1}$	V_o (ml/mol)
Methylstearate	16.4	8.3	11.3	352.0
Ethylstearate	16.6	8.2	10.8	370.5
n-Butylstearate	17.1	8.0	10.1	405.3
n-Octylstearate	18.5	6.8	8.9	475.5
n-Hexadecylstearate	20.8	5.4	6.3	614.2

[*]Equation (1): correlation coefficient better than 0.9998 and
equation (3): correlation coefficient better than 0.9990.

E_a is obviously expected to increase with increasing molecular dimensions, a trend reflected in Fig. 3, where the experimental data for the stearates are plotted as a function of the number of chain atoms Z. Some additional E_a results for linear aliphatic esters of shorter chain length, calculated from literature data (Landolt-Börnstein, 1969) are also plotted in the Figure.

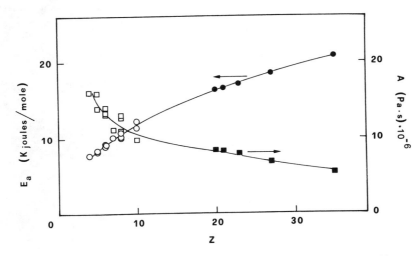

Fig. 3.   Viscous flow parameters E_a and A (see equation (1)) as a function of the number of chain atoms Z. Black symbols: stearates (this work); open symbols: esters (from literature data).

Kauzmann and Eyring (1940) found that E_a for linear paraffins is approximately 1/4 of the heat of vaporization ΔH_v at low values of Z, whereas, as the number of chain atoms increases, E_a and ΔH_v diverge, E_a tending to a plateau value for Z>30 chain atoms. This was taken as evidence that the flow units, in long chain molecules, are constituted by segments of such a length. The E_a values in Fig. 3 increase from about 8 KJ/mol for low molecular weight esters to about 20 KJ/mol for the stearate of n-Hexadecyl alcohol. This behaviour well compares with Kauzmann's results on hydrocarbons, although there could be a levelling off of the E_a vs. Z curve for esters only at a chain length higher than Z=35. Figure 3 also shows the preexponential term A of equation (1) which regularly decreases with increasing Z.

A different way to plot viscosity data, firstly proposed by Batschinski (1913), has been recently applied by Hildebrand and Lamoreaux (1972) to several liquids. According to the above authors, viscosity is not a direct function of temperature but of the excess of the liquid molar volume V over the volume V_o at which the molecules are so closely crowded that they cannot undergo viscous flow:

$$\phi = 1/\eta = B(V-V_o)/V_o \qquad\qquad (3)$$

where ϕ is the fluidity and B is a constant for a given liquid and, according to Alder and Hildebrand (1973), "it depends inversely upon the capacity of molecules to absorb the externally imposed momentum of viscous flow by reason of their mass, flexibility, softness or inertia of rotation". The primary effect of temperature (and pressure) is simply to determine the magnitude of $V-V_o$.

The viscosity data of the stearates investigated in this work, plotted according to Hildebrand, are shown in Fig. 4. The data bend away from the drawn lines when the fluidity is low, as found for other liquids in the proximity of the melting temperature (Hildebrand and Lamoreaux, 1973). However, the plots are linear at high fluidities, i.e. at temperatures higher than $T_m+100°C$, so that the constants B and V_0 can be obtained using equation (3) and the results are collected in Table 2.

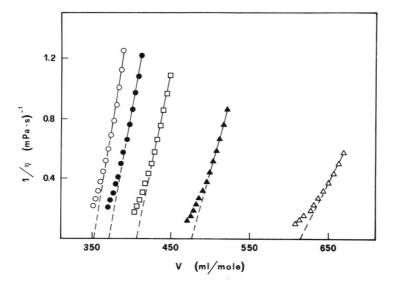

Fig. 4.   Fluidity as a function of molar volume for :
○ Methyl-; ● Ethyl-; □ n-Butyl-; ▲ n-Octyl-
and △ n-Hexadecylstearate.

The dependence of B on molecular dimensions (number of chain atoms Z) is plotted in Fig. 5 for the stearates together with alkanes data from Hildebrand and Lamoreaux (1972) and with values calculated from viscosity and density literature data (Landolt-Börnstein, 1969) concerning linear aliphatic esters of shorter chain length. The Figure shows that the fluidity parameter of hydrocarbons and esters follows a similar linear dependence on the number of chain atoms, suggesting that the substitution of a CH_2-CH_2 group with CO-O in a linear aliphatic carbon chain modifies very little the temperature and chain length dependence of viscosity.

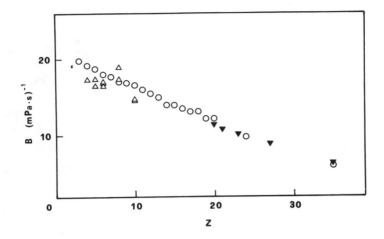

Fig. 5. Fluidity parameter B as a function of number of chain
atoms Z: ▼ stearates (this work); ○ alkanes (Hildebrand
and Lamoreaux, 1972) and △ esters (calculated from
literature data).

Glass Transition Temperature of the PVC-stearates Mixtures

The glass transition temperature of polymers is usually strongly depressed by low
molecular weight substances that interact with the macromolecular chains. When the
solubility of the diluent in the polymer is limited, the glass transition
temperature of the diluted polymer remains constant at any concentration higher
than that corresponding to the solubility limit, as shown by Onu, Legras and
Mercier (1976) for the system Polycarbonate-plasticizers. The glass transition
temperatures of the PVC-stearate systems examined are plotted in Fig. 6 as a
function of the total amount of ester, R, expressed as g ester/g PVC. With
increasing R, a plateau value is reached for all mixtures, a result which clearly
indicates limited solubility of the stearates in the polymer. The plateau values
of T_g, collected in Table 3, are closely related to the solubility R_o of the esters
in PVC; in fact, the higher the solubility the lower the limit value of T_g. The
critical composition R_o was evaluated as previously described (Ceccorulli and co-
workers, 1981) from the DSC melting endotherms of the excess undissolved diluent.
The T_g plateau value depends on the molecular structure of the stearate, lying as
much as 26°C below the pure polymer T_g for Methylstearate and only 2°C for n-Hexadecylstearate. The remaining esters give intermediate values. The T_g-composition
curve for the system PVC-Stearic acid, also plotted in Fig. 6, shows that Stearic
acid - a widely used lubricant for PVC - is more effective than n-Hexadecylstearate
in lowering T_g.

The dependence of the limit value of T_g on molecular structure is illustrated in
Fig. 7, where the plateau value of T_g is plotted as a function of the reciprocal
number of carbon atoms in the linear alcoholic residue, 1/n. The decrease of T_g
appears to be rather regular and it seems possible to carry out an extrapolation

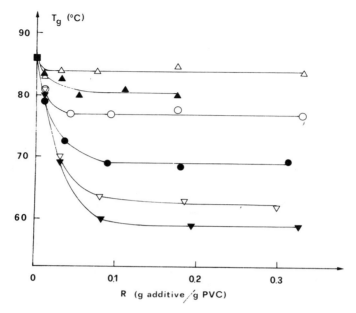

Fig. 6. Glass transition temperature as a function of additive content:
▼ PVC-Methylstearate; ▽ PVC-Ethylstearate; ● PVC-n-Butylstearate;
○ PVC-n-Octylstearate; △ PVC-n-Hexadecylstearate; ▲ PVC-Stearic
acid and ■ pure PVC.

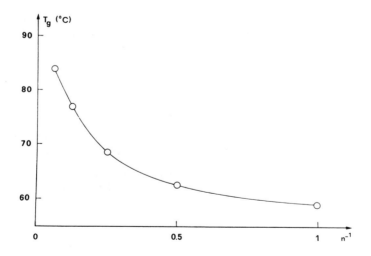

Fig. 7. Glass transition temperature (plateau value) of the PVC-
stearate systems as a function of the reciprocal number
of carbon atoms in the alcoholic residue of the ester.

to $n \to \infty$ $(1/n \to 0)$, i.e. to a stearate of an infinitely long – and therefore completely insoluble – alcohol. This would lead to an extrapolated T_g value of about $90°C$ which satisfactorily compares with the glass transition temperature of pure PVC $(86°C)$.

In order to investigate the dependence of the plateau value of T_g on the compatibility of the various stearates with PVC, the glass transition temperatures are plotted in Fig. 8 as a function of the ester volume fraction ϕ_1. The points represent the experimental data and are magnified in the lower insert; ϕ_1 is calculated at the solubility limit assuming volume additivity, as follows:

$$\phi_1 = (R_o/\rho_1) / (R_o/\rho_1 + 1/\rho_2) \tag{4}$$

where ρ_1 and ρ_2 are the densities of diluent and polymer respectively, at the temperature corresponding to the glass transition of the mixture. Some additional data for PVC-ester mixtures processed at $150°C$ (open triangles) are also included in Fig. 8. It is clear that the points corresponding to different diluents all lay on the same curve which is almost a straight line in the range of composition experimentally explored.

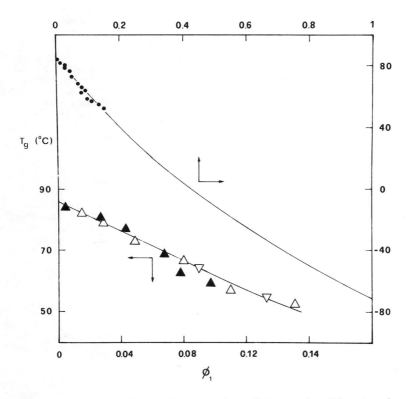

Fig. 8.   Dependence of the plateau value of T_g on the diluent volume fraction ϕ_1 for PVC-stearate mixtures processed at $90°C$ (▲) and at 150 $°C$ ($\triangle,\nabla$). Solid lines represent equation (6).

Keeping in mind that the slope of the T_g vs. composition curve for any polymer-diluent system mainly depends on the free volume parameters of the diluent, i.e. its glass transition temperature T_{g1} and free volume expansion coefficient $\Delta\alpha_1$ (Kelley and Bueche, 1961), the results of Fig. 8 suggest that all the esters examined have similar free volume characteristics.

The temperature dependence of the free volume of any liquid is usually described by the equation:

$$f = f_g + \Delta\alpha \cdot (T - T_g) \qquad (5)$$

where f_g is the fractional free volume at T_g, usually taken as 0.025 (Ferry, 1970), and the free volume expansion coefficient $\Delta\alpha$ is approximately given by $\alpha_1 - \alpha_g$, i.e. by the difference between the thermal expansion coefficient of the liquid and the glass respectively. For a homogeneous solution of polymer and diluent of volume fraction ϕ_2 and ϕ_1 respectively, assuming that the free volumes are additive and that T_g is an iso-free volume state, the glass transition temperature of the mixture is expressed by Braun and Kovacs (1965) in the form:

$$T_g = T_{g2} - \{ \phi_1\Delta\alpha_1 \cdot (T_{g2} - T_{g1})/(\phi_1\Delta\alpha_1 + \phi_2\Delta\alpha_2) \} \qquad (6)$$

where T_g, T_{g2} and T_{g1} (in $^\circ K$) are the glass transition temperatures of the system, pure polymer and diluent respectively and $\Delta\alpha_2$ and $\Delta\alpha_1$ are the free volume expansion coefficients of polymer and diluent.

In order to apply equation (6) to the PVC-ester systems, numerical values for the polymer and diluent parameters are required. While $\Delta\alpha_2$ can be found in the literature, no T_{g1} and $\Delta\alpha_1$ data are known for the stearates studied in the present work. Several attempts to quench the esters from the liquid to the glassy state failed, so that the glass transition temperatures had to be estimated from the approximate relationship $T_g \cong (2/3) T_m$. Being the melting temperatures of the esters rather close to each other (see Table 1), a common value of T_{g1} was used (T_{g1} =202 $^\circ K$). The solid lines in Fig. 8 represent equation (6) calculated using the experimental values T_{g2} = 359$^\circ K$ and $\Delta\alpha_2$ = $3.4 \cdot 10^{-4}$$^\circ C^{-1}$ (Pezzin, Omacini and Zilio-Grandi, 1968); $\Delta\alpha_1$ was taken as an adjustable parameter, whose value in the present case was $5.6 \cdot 10^{-4}$$^\circ C^{-1}$. The calculated T_g versus ϕ_1 curve over the entire composition range is drawn in the upper part of the Figure and the lower magnification shows that the fit with the experimental data is satisfactory over the composition range explored. It is interesting to note that the value assigned to $\Delta\alpha_1$, when compared with the measured volume expansion coefficient of the esters in the liquid state (average value α_1= $8.9 \cdot 10^{-4}$ $^\circ C^{-1}$) gives the quite reasonable value of $3.3 \cdot 10^{-4}$ $^\circ C^{-1}$ for their expansion coefficient in the glassy state.

Flory-Huggins Interaction Parameter

The solubility of the stearates in PVC decreases from 6.5% to 0.3% with increasing chain length, as shown in Table 3. Changes in solubility should reflect differences in type, intensity and/or frequency of the intermolecular forces which develop between polymer and diluent. In the attempt to quantify the interaction forces, the experimental critical composition R_o can be used to evaluate the Flory-Huggins thermodynamic interaction parameter χ. A widely accepted model (Alfrey and co-workers, 1949; Manson, Iobst and Acosta, 1974) considers PVC, a polymer with a low degree of crystallinity, as a macromolecular network in which the microcrystals act as physical crosslinks. According to this model and assuming that R_o represents the equilibrium swelling composition at $90°C$, the equation by Flory and Rehner (1943) can be used to obtain χ:

$$\ln (1-\phi_2) + \phi_2 + \chi\phi_2^2 + (\rho_2 V_1/M_c) \cdot (\phi_2^{1/3} - \phi_2/2) = 0 \qquad (7)$$

where V_1 is the diluent molar volume and M_c the molecular weight between crosslinks in the polymer. As regards M_c, an average value of 2500 was used in the calculations, according to previously published data (Pezzin, Ajroldi and Garbuglio, 1969; Manson, Iobst and Acosta, 1974).

Assuming volume additivity, the polymer volume fraction ϕ_2 at the solubility limit was calculated and the application of equation (7) yielded the χ values reported in Table 3. As expected for poorly compatible systems, the values of χ are

TABLE 3   Equilibrium Parameters at $90°C$ for PVC-diluent Systems

Diluent	R_o (g/g)	T_g^* (°C)	χ
Stearic acid	0.017	80.5	2.7
Methylstearate	0.065	59.0	1.6
Ethylstearate	0.051	62.5	1.8
n-Butylstearate	0.044	68.5	1.9
n-Octylstearate	0.027	77.0	2.2
n-Hexadecylstearate	0.003	84.0	4.2

*Plateau value

considerably high and increase regularly with decreasing solubility, i.e. with increasing chain length of the stearates. This result supports the hypothesis that the PVC-ester interaction takes place mainly between the chlorine atoms of the polymer and the C=O groups of the ester, as previously suggested by Tabb and Koenig (1975) in an infrared investigation on PVC-phthalate systems. The lengthening of the aliphatic chain along the homologous series of stearates clearly results in a dilution of the ester groups and therefore in a decrease of the PVC-ester interaction sites. As far as Stearic acid is concerned, its χ value lies between those of n-Octyl and n-Hexadecylstearate; the same occurs to the plateau value of T_g. Apparently, Stearic acid behaves like a linear ester of relatively high molecular weight, a result that can be explained assuming that the acid is present in the polymer as a dimer (Nagy and co-workers, 1974).

In conclusion, the main results of this work can be summarized as follows:
a) the viscous flow parameters of the series of linear aliphatic stearates show a chain length dependence similar to that previously found for linear alkanes;
b) the decrease of the glass transition temperature of PVC brought about by the dilution with the esters is satisfactorily described by the free volume theory by means of a single equation provided the esters' T_g is taken as $202°K$ and the free volume expansion coefficient as $5.6 \cdot 10^{-4}°C^{-1}$;
c) a definite equilibrium solubility in PVC at $90°C$ exists for the stearates as well as for Stearic acid. The solubility decreases regularly with increasing

chain length and in Flory-Huggins terms this corresponds to a regular increase of the thermodynamic interaction parameter χ.

ACKNOWLEDGEMENT

This work was supported by C.N.R. (Progetto Finalizzato Chimica Fine e Secondaria).

REFERENCES

Alder, B.J., and J. H. Hildebrand (1973). Activation energy: not involved in transport processes in liquids. I&EC Fundamentals, 12, 387-388.
Alfrey, T. Jr., N. Wiederhorn, R. Stein, and A. Tobolsky (1949). Plasticized Polyvinylchloride. J. Colloid Sci., 4, 211-227.
Anagnostopoulos, C. E., and A. Y. Koran (1962). Polymer diluent interactions. II. Polyvinylchloride-diluent interactions. J. Polymer Sci., 57, 1-11.
Batschinski, A. J. (1913). Untersuchungen über die innere Reibung der flüssigkeiten. I. Z. Physik. Chem., 84, 643-706.
Bondi, A. (1956). Theory of viscosity. In F. R. Eirich (Ed.), Rheology. Academic Press, New York. Chap. 7.
Braun, G., and A. J. Kovacs (1965). Variation de la température de transition vitreuse dans les systèmes binaires à répartition statistique. In J. A. Prins (Ed.), Physics of Non-Crystalline Solids. North-Holland, Amsterdam. Part 3, pp. 303-319.
Ceccorulli, G., F. Manescalchi, and M. Pizzoli (1975). Thermal behaviour of Nylon 8. Makromol. Chem., 176, 1163-1171.
Ceccorulli, G., M. Pizzoli, and G. Pezzin (1977). Effect of thermal history on T_g and corresponding c_p changes in PVC of different stereoregularities. J. Macromol. Sci. (Phys.), B14, 499-510.
Ceccorulli, G., M. Pizzoli, M. Scandola, G. Pezzin, and G. Crose (1981). Interaction of partially soluble additives with Polyvinylchloride: solubility of stearates at high temperature. J. Macromol. Sci. (Phys.), in press.
Ceccorulli, G., M. Scandola, and G. Pezzin (1977). Calorimetric investigation of some elastin-solvent systems. Biopolymers, 16, 1505-1512.
Doty, P., and H. S. Zable (1946). Determination of polymer-liquid interaction by swelling measurements. J. Polymer Sci., 1, 90-101.
Ewell, R. H., and H. Eyring (1937). Theory of the viscosity of liquids as a function of temperature and pressure. J. Chem. Phys., 5, 726-736.
Ferry, J. D. (1970). Viscoelastic Properties of Polymers. Wiley, New York. Chap. 11.
Flory, P. J., and J. Rehner (1943). Statistical mechanics of cross-linked polymer networks. II. Swelling. J. Chem. Phys., 11, 521-526.
Frenkel, J. (1926). Über die Wärmebewegung in fest und flüssigen Körpern. Z. Physik., 35, 652-669.
Glasstone, S., K. J. Laidler, and H. Eyring (1941). The Theory of Rate Processes. Mc Graw-Hill, New York.
Hildebrand, J. M., and R. M. Lamoreaux (1972). Fluidity: a general theory. Proc. Nat. Acad. Sci. USA, 69, 3428-3431.
Hildebrand, J. M., and R. H. Lamoreaux (1973). Fluidity and liquid structure. J. Phys. Chem., 77, 1471-1473.
Kauzmann, W., and H. Eyring (1940). The viscous flow of large molecules. J. Am. Chem. Soc., 62, 3113-3125.
Kelley, F. N., and F. Bueche (1961). Viscosity and glass temperature relations for polymer-diluent systems. J. Polymer Sci., 50, 549-556.
Landolt-Börnstein (1969). Zahlenwerte und Funktionen. 6.Auflage, Band II, Teil 5/a. Springer-Verlag, Berlin.

Manson, J. A., S. A. Iobst, and R. Acosta (1974). Thermomechanical behavior and
    structure of Polyvinylchloride. J. Macromol. Sci. (Phys.), B9, 301-320.
Nagy, J., S. Ferenczi-Gresz, R. Farkas, T. T. Nagy, and B. Pukanszky (1974).
    Dielektrizitätsspektroskopische Untersuchung von PVC-Gleitmittel-Systemen
    unterhalb der Glastemperatur. Plaste und Kautschuk, 21, 919-921.
Onu, A., R. Legras, and J. P. Mercier (1976). Phase equilibrium and glass
    transition temperatures in plasticized amorphous bisphenol-A Polycarbonate.
    J. Polymer. Sci., 14, 1187-1199.
Pezzin, G., G. Ajroldi, and C. Garbuglio (1969). Influence of molecular weight on
    some rheological properties of plasticized polyvinylchloride. Rheol. Acta, 8,
    304-311.
Pezzin, G., A. Omacini, and F. Zilio-Grandi. (1968). La transizione vetrosa dei
    sistemi cloruro di polivinile-diluente. Chim. Ind. (Milan), 50, 309-313.
Tabb, D. L., and J. L. Koenig (1975). Fourier transform infrared study of
    plasticized and unplasticized Polyvinylchloride. Macromolecules, 8, 929-934.

EVALUATION OF ETHYLENE COPOLYMERS AS POUR DEPRESSANTS

Richard F. Miller
ARCO Performance Chemicals Company, Inc.
1500 Market Street
Philadelphia, Pennsylvania 19101

ABSTRACT

The precipitation of paraffin waxes occurring at low temperatures during the transportation of distillate fuel oils reduces the flow rate of these fluids and creates problems which plague the petroleum industry. Pour depressants are used as additives to reduce the size and shape of the crystalline precipitate and to inhibit the formation of cage-like structures. In this investigation, attempts have been made to relate the solubility parameter of the additive to its perform-ance in selected fuel oils. Polyethylene is not particularly useful as a pour depressant; however, ethylene copolymers having solubility parameters in the 8.6 H range were found to be effective.

KEYWORDS

Solubility parameters, ethylene copolymers, pour point depression, distillate fuel oil additives.

INTRODUCTION

The precipitation of paraffin wax during the transportation of distillate fuel oil at low temperatures reduce the flow rate of these fluids and creates a problem which plagues the petroleum industry. These lamellar wax precipitates form cage-like structures which impede liquid flow (Pass, 1967).

Pour depressants are used as additives which reduce the size of the crystalline precipitate and inhibit the formation of cage-like structures (Gavlin, 1953; Holder, 1965; Koch, 1955).

While several investigators have attempted to relate the structure of these additives with their effectiveness, no generally acceptable mechanism for pour depressant action has been devised (Dimitroff, 1969; Lorensen, 1962; Tiedje, 1961). An attempt has been made in this investigation to relate solubility parameters of the additive and oil with the performance of select useful polymeric pour point depressants.

SOLUBILITY PARAMETER CONCEPTS

The concept of solubility parameters is an attempt to quantify the old rule of thumb, "like dissolves like," based on the following considerations:

1. Transfer from the liquid to the gaseous state requires overcoming an inter-
    action energy of $z\varepsilon_j/2$ per molecule and consequently, $N_L z\varepsilon_j/2$ per mole. This
    is equal to the negative internal mole energy of vaporization ($\Delta\varepsilon$). The term
    ε_j is the energy per bond. One molecule has z neighbors and the corresponding
    quantity related to the molar volume V^m is called the cohesive energy density
    (CED) as shown in equation 1.

$$CED_j = \frac{\Delta\varepsilon_j}{V^m_j} \qquad = \qquad \frac{-.05\ N_L\varepsilon_j z}{V^m_j} \qquad\qquad (1)$$

The solubility parameter is defined as the square root of the cohesive energy
density as shown in equation 2.

$$\delta_j \quad = \quad (CED_j)^{\frac{1}{2}} \qquad\qquad (2)$$

2. Interaction energies ε are related to each other in the following manner.
    Mixing solvent 1 and polymer 2 produces two solvent-polymer (1-2) bonds for
    every broken solvent-solvent (1-1) and polymer polymer (2-2) bond. The change
    in interaction energy during the mixing process is shown in equations 3 and 4.

$$\Delta\varepsilon = \varepsilon_{12} - 0.5\ (\ \varepsilon_{11} + \varepsilon_{12}\ ) \qquad\qquad (3)$$

$$-2\ \Delta\varepsilon = (\varepsilon^{\frac{1}{2}}_{11})^2 - 2\ \varepsilon_{12} + (\varepsilon_{22}{}^{\frac{1}{2}})^2 \qquad\qquad (4)$$

The interaction energy of two different spherical molecules due to dispersion forces
is equal to the mutual interaction energies of the molecules themselves, i.e.

$$\varepsilon_{12} = -(\ \varepsilon_{11}\ \varepsilon_{22})^{\frac{1}{2}} \qquad\qquad (5)$$

which gives $\qquad\qquad \Delta\varepsilon = -0.5\ (\varepsilon_{11}{}^{\frac{1}{2}} - \varepsilon_{22}{}^{\frac{1}{2}})^2$

Assuming equal molar volumes of solvent and polymer monomeric unit gives:

$$\frac{0.5\ z\ N_L\Delta\varepsilon \quad = \quad -0.5\ (\ \delta_1 - \delta_2\ )^2}{V^m} \qquad\qquad (6)$$

The difference in solubility parameter values thus yields a measure of the interaction between solvent and solute with respect to the mutual interactions between like
components. If ε_{11} and/or $\varepsilon_{22} = \varepsilon_{12}$ then there will be practically no interaction
between solvent and solute. With equal interactions between 1-1, 2-2, and 1-2,
then $\delta_1 - \delta_2 = 0$ at which it is still possible to have mixing. The experimentally obtained maximum difference varies according to the polarity of the solvent. In
practice, as a first approximation and in the absence of strong interactions, such
as hydrogen bonding, solubility or compatibility on the molecular level can be
expected if $\delta_1 - \delta_2$ is less than 1.8 H (Seymour, 1981).

δ for a nonvolatile material of known structure may be determined by the use of the Small equation shown below,

$$\delta_2 = \frac{D\Sigma G}{M} \qquad (7)$$

wherein values of G, the molar attraction constants are summed over the structural configuration of the repeating unit in the polymer chain, with respect to the molecular weight (M) and density (D).

A DISCUSSION

Since copolymers of ethylene with 17-40% of vinyl acetate have solubility parameter values similar to those of distillate fuels, twelve of these copolymers were chosen for use as potential pour point depressants. Neither polyethylene nor polyvinyl acetate were particularly useful for this application.

The number average molecular weight of these copolymers varied from 1700 to 4700. However, these values were all above the threshold of molecular weight required for depressant activity and below the size that would be highly viscous. Determination of hydroxyl content showed that all copolymer samples contained less than 1 percent of vinyl alcohol repeating units and, hence, this was not considered to be a factor in determining the effectiveness of these copolymers.

The solubility parameter values as calculated from Small's formula are also shown in Table 1. The experimental values for solubility parameter of these additives are also shown in this table.

Some of the discrepancy between the calculated and experimental values are believed to be related to the degree of branching of the copolymers. The extent of chain branching was also determined by ^{13}C NMR. Corrections of the calculated Δ values for branching, provided new values that agreed better with the experimental values.

Experimental values attained by titration and viscosity measurements were used to predict performance of these additives as pour point depressants.

As shown in Table 2, five samples were soluble in and were effective pour depressants at concentrations of 1000 ppm when added to a mid-continent oil having a pour point of -10°F.

As shown in Table 3, the copolymer samples at concentrations of 500 and 1000 ppm were also effective in lowering the pour point of a #2 fuel oil. However, unlike the effects shown in Table 2, these additives were also effective in concentrations of 500 ppm. Chlorinated polyethylene waxes with appropriate solubility parameter values were also effective as pour point depressants.

EXPERIMENTAL

The copolymers were prepared by heating ethylene and vinyl acetate under high pressure in the presence of benzoyl peroxide.

The number average molecular weights were determined by GPC (Matsuda, 1977). The presence of hydroxyl groups and the extent of chain branching were determined by ^{13}C NMR (Wu, 1974).

TABLE 1   Solubility Parameters and Molecular Weight Values
for Ethylene/Vinyl Acetate Copolymer

Sample No.	Ratio Et/VAc	δ_{CALC}	δ_{EXP}	δ_{CORR}
1	85/15	8.2	8.7	8.7
2	70/30	8.5	9.0	9.2
3	60/40	8.6	9.5	9.2
4	63/37	8.6	8.6	---
5	68/32	8.5	9.2	9.1
6	69/31	8.5	8.8	8.8
7	74/26	8.4	8.7	---
8	77/23	8.3	9.0	9.1
9	74/26	8.4	9.3	9.2
10	66/34	8.5	8.7	---
11	60/40	8.6	9.25	9.2
12	60/40	8.6	8.6	---

TABLE 2   Fuel A: Mid-Continent Area, Untreated Pour Point - 10°F
(δ = 9.0 H)

Polymer Sample No. Treatment (ppm)	1	3	4	6	10
250	-15	-10	-15	-10	-15
500	-15	-10	-15	-10	-30
1000	-35	-15	-60	-45	-45
2000	-15	-15	-60	-20	-60

TABLE 3 Fuel B: No. 2 Fuel Oil, Untreated Pour, +5°F
(δ = 8.9 H)

Polymer Sample No. Treatment (ppm)	1	2	4	6	7	8	9	10	12
500	-15	-15	-25	-25	-20	-10	-10	-20	-10
1000	-25	-15	-40	-20	-35	-15	-10	-40	-25

The solubility parameter values were determined by titrating with more polar and less polar solvents and by noting optimum intrinsic viscosities. In the first methods, 1.0 g. of polymer was dissolved in a known volume of toluene and a measured volume of methanol was added until a stable turbid system was produced. This procedure was repeated with n-heptane. The solubility parameter of the copolymer was determined from the summation of the product of the volume percent of toluene and its solubility parameter and that of the non-solvent and its solubility parameter. The average value for these titrations was used to obtain the upper and lower turbidity limits and the mean of these two values was used as the solubility parameter of the additive.

In the intrinsic viscosity method, 1.0 g. of additive was dissolved in four different solvents and the solubility parameters of the solvents were plotted against the intrinsic viscosity. The solubility parameter was considered as the value at the maximum intrinsic viscosity. Both techniques gave similar results.

The pour point of the oils were determined by ASTM method D-97.

CONCLUSIONS

While neither polyethylene nor polyvinyl acetate were found to be useful as pour point depressants for fuel oils, ethylene copolymers with average molecular weight values of 1700-4700 and having solubility parameters in the 8.6 H range were found to be effective pour point depressants. Copolymers outside this range exhibited varying degrees of performance. By increasing the additive dosage above that required for measurable pour depression, smaller crystals and better low temperature flow properties were not observed as previously reported (Holder, 1965).

REFERENCES

Dimitroff, E. and Dietzmann, H. (1969). Am. Chem. Soc. Div. Petrol. Chem. Prepr., 14, B-132.
Gavlin, G., Swaine, E. A. and Jones, S. P. (1953). Ind. Eng. Chem., 45 (10), 2327.
Holder, G.A. and Winkler, J. (1965). J. Inst. Pet. (London), 51 (499), 228.
Koch, E. (1955). Erdolund Kohle, 8, 793.
Lorensen, L. E. (1962a). Am. Chem. Soc. Div. Petrol. Chem. Prepr., 7 (4), B-61.
Lorensen, L. E. and Hewett, W. A. (1962h). Am. Chem. Soc. Div. Petrol. Chem. Prepr. 7 (4), B-71.
Matsuda, H. and Co-workers (1977). Polym. J., 9 (6), 527-535.
Pass, F. J. Csoklich, Ch. and Wastl, K. (1967). World Petrol. Congr. Proc. 7th, 8, 129.

Seymour, R. B. and Carraher , C. (1981). Polymer Chemistry: An Introduction, Marcel
    Dekker, Inc., Chap. 3.2.
Tiedje, J. L. and Hollyday, W. C. (1961). Hydrocarbon Process, Pet. Refiner, 40 (8),
    111.
Wu, T.K., Ovenall, D. W. and Reddy, G. S. (1974). J. Poly. Sci., 12, 901-911.

PREPARATION AND PROPERTIES OF CELLULOSE BLENDS IN DIMETHYLSULFOXIDE SOLUTION

Raymond B. Seymour[1], Earl L. Johnson[2], and
G. Allan Stahl[3]

University of Houston, Department of Chemistry
Houston, Texas 77004

ABSTRACT

Solutions of cellulose where prepared via the _in situ_ formation of cellulose methylols by heating mixtures of dimethylsulfoxide (DMSO), paraformaldehyde (PF) and Whatman Filter paper. These 1.0 percent stock solutions of cellulose were used to prepare homogeneous blends of other DMSO soluble polymers with cellulose. Clear homogeneous solutions of polyacrylonitrile (PAN), polyvinyl-pyrrolidone (PVP) and Poly(vinyl alcohol) (PVA) were prepared as 0.3, 1.0 and 3.0 percent solutions in 1.0 percent solutions of cellulose dissolved in DMSO:PF. These solutions were characterized using Brookfield viscometry and used to cast cellulose films containing various amounts of PAN, PVP and PVA. These films were characterized physically and found to exhibit tensile strengths ranging between 130 and 540 kg/cm^2 as a function of the cellulose polyblend composition. The films were characterized morphologically via scanning electron microscopy (SEM) and the cellulose/PVA films were found to be porous. The average pore diameter varied between 5.0 and 80.0 microns as a function of film composition.

[1]Present address: Department of Polymer Science
University of Southern Mississippi
Hattiesburg, Mississippi 39401

[2]Present address: E. I. DuPont de Nemours and Co.
Wilmington, Delaware 19898

[3]Present address: Phillips Petroleum Company
Bartlesville, Oklahoma 74004

KEYWORDS

Cellulose methylols, DMSO, paraformaldehyde, tensile strength, cellulose
polyblend, scanning electron microscopy, film pore diameter.

INTRODUCTION

Louis Menard's 1846 discovery of collodion, a solution of cellulose nitrate
in a mixture of ethanol and ethyl ether, was the precursor of celluloid,
rayon, and the cellulosic lacquers. Subsequently, cellulose was dissolved
in sulfuric acid and aqueous solutions of sodium hydroxide, zinc chloride,
and cupric ammonium hydroxide (Schweitzer's reagent). The solutions were used
in the preparation of parchment paper, vulcanized fiber, and mercerized
cotton. Although very useful, these solutions are not satisfactory for use
in the preparation of cellulose esters and ethers because of the presence of
water (Ott, Spurlin, and Grafflin, 1971). Further, blends of cellulose and
many commercial synthetic polymers cannot be prepared because of the limited
solubility of these polymers in water.

In this investigation, cellulose was dissolved in dimethylsulfoxide (DMSO) in
the presence of formaldehyde (generated in situ via thermal decomposition of
paraformaldehyde (PF) in DMSO) via the formation of cellulose methylols (Johnson,
Nicholson, Haigh, 1975; Seymour, Johnson, 1976). These clear viscose solutions
were blended with other polymers to yield cellulose polyblends which were used
to prepare cellulosic films of varying tensile strengths and porosities. The
film forming qualities of these cellulose polyblend solutions were shown to be
a direct consequence of the simularities in solubility paramater (δ) displayed
by cellulose, DMSO, PF and the blended polymer. Stock solutions of cellulose
dissolved in DMSO:PF were blended with poly(vinyl alcohol) (PVA), polyvinyl-
pyrrolidone (PVP), polyacrylonitrile (PAN), polystyrene (PS), poly(methyl
methacrylate) (PMMA), poly(vinyl acetate) (PVOAc), poly(ethylene terephthalate)
(Dacron) and Nylon-66 were prepared and used to cast cellulosic films which
were inspected using scanning electron microscopy (SEM) and evaluated physically
by tensile testing.

EXPERIMENTAL

Stock solutions of cellulose were prepared by placing 20.0 g of cellulose
(Whatman filter paper), 40.0 g of paraformaldehyde (PF) and 2,000 g of DMSO in
a closed glass container and heating this mixture for 24 hours at 75°C with
occasional shaking. The water clear viscose cellulose solutions (1.0 percent
cellulose) exhibited viscosities between 3,000 and 4,000 centipoise. Similar
solutions were made using other sources of cellulose, e.g.: cotton fabric and
various types of paper. In this investigation Whatman filter paper was used
as the exclusive source of cellulose.

Cellulose polyblends were prepared by adding various amounts of PVA, PAN, PS,
PMMA, PVOAc, PVP, Dacron and nylon-66 to 100 g of the 1.0 percent stock
solution of cellulose dissolved in DMSO:PF. The resulting solutions were
agitated continuously for 24 hours and the solubility of the blended polymer
evaluated by inspection. Brookfield viscosities of the cellulose polyblend
solutions were determined at 27°C using standard Brookfield techniques (spindle
No. 6 at 50 rpms).

Cellulosic films were prepared from the homogeneous cellulose polyblends by casting them onto 2.5 x 6.0 cm glass slides at 50°C for 18 hours after-which the temperature was raised to 75°C for an additional 18 hour period. The films were then stored under vacuum over P_2O_5 for 24 hours, and then placed in a 90 percent relative humidity enviornment for 24 hours. It was noted that storage of the films in high humidity environments had a plasticizing effect and the films were then easily removed from the glass slide.

Water absorption characteristics of the cellulosic films were determined by first drying the films (1.0 x 4.0 cm) over P_2O_5 under vaccum at room temperature for three days. The films were then weighed and then stored in a 62.5 percent relative humidity chamber for three days. The films were then reweighed and the percent weight increase noted and correlated with the composition of the film.

Tensile test were preformed on 1.0 x 4.0 cm test samples. A Lebow Model No. 3397 load cell and a MTS mechanical tester were used. The test samples were stabilized in a 50 percent relative humidity environment at 21°C for 72 hours prior to testing per ASTM specifications for semirigid plastics at a 5.0 cm/min load rate.

The samples were prepared for SEM analysis by coating with Au/Po on a Technics Corporation Hummer. They were viewed at 71,000x and 35,500x on a Cambridge Stereoscan Model S410 Scanning Electron Microscope.

Dacron[R], Nylon-66[R], and Elvanol[R] commercial polymers were used in this study. PAN, PS, PMMA, PVOAc, and PVP were prepared by heating the monomer (10g) in a diluent (50 ml) with azobisisobutronitrile (DuPont Vazo-64) (0.10g) at 60°C. The diluents were benzene, hexane, hexane, cyclohexane, and benzene respective-ly. The monomers were reagent grade. The polymers were isolated by filtration and dried before use to ascertain quantitative conversion.

RESULTS AND DISCUSSION

As shown in Table I, mixtures of paraformaldehyde and cellulose (in a 10:1 molar ratio based on glucose and formaldehyde) are not soluble in other polar aprotic solvents such as N,N-dimethylformamide (DMF) and hexamethylphosphora-mide (HMPA) nor in polar protic solvents at 75°C. It should also be noted that the cellulose and formaldehyde mixture is not soluble in anhydrous DMSO. Thus, the presence of methanediol in the solution has been assumed. (Johnson and Nicholson, 1975; Chapman and King, 1964), in explaining cellulose solubility in DMSO/paraformaldehyde when traces of water present.

Paraformaldehyde-cellulose (10:1 molar ratio) mixtures in DMSO produced homogeneous solutions or gels. These were all found to be miscible in all proportions with DMF, DMAc, HMPA, and pyridine. However, cellulose was precipitated from these homogeneous solutions and gels when dioxane, alcohol, water, or solvents with similar solubility parameters were added. Brittle films and weak fibers can be obtained by solvent evaporation, precipitation in water or ethanol, or wet spinning in water or alcohol.

TABLE I

Solubility of Cellulose-Paraformaldehyde Mixture in Selected Solvents at $75^{\circ}C$[a]

Solvent	Hildebrand solubility parameters δ, H	Solubility[b]
Pyridine	10.7	-
N-Methyl-2-pyrrolidone	11.3	-
Acetonitrile	11.9	-
Benzyl alcohol	12.1	-
DMF	12.1	-
HMPA	12.	-
Dimethylacetamide (DMAc)	12.2	-
Nitromethane	12.7	-
Sulfolane (tetrahydrothiophene sulfane)	12.	-
Maleic anhydride	13.6	-
DMSO	12.0	+

a – Each mixture consisted of 0.5 g cellulose, 1.0 g paraformaldehyde, and 98.5 g solvent.

b – + Indicates Solubility; – Indicates Insolubility

DMSO is known to form strong bonds with various alcohols (Chapman and King, 1964a; Rader, 1966; and Jacob, Rosenbaum, and Wood, 1971). Those bonds have in turn been postulated to form six membered complexes as shown in Figure 1A. Knowing of these interactions, other workers have prepared similar complexes of DMSO with dextran and oligoglucose (Chapman and King, 1964a, 1964b; Casu Reggiani, Gallo, and Vigevani, 1964, 1965, 1966). These workers employed PMR to demonstrate the presence of complexes, and they proposed structures similar to those in Figure 1A.

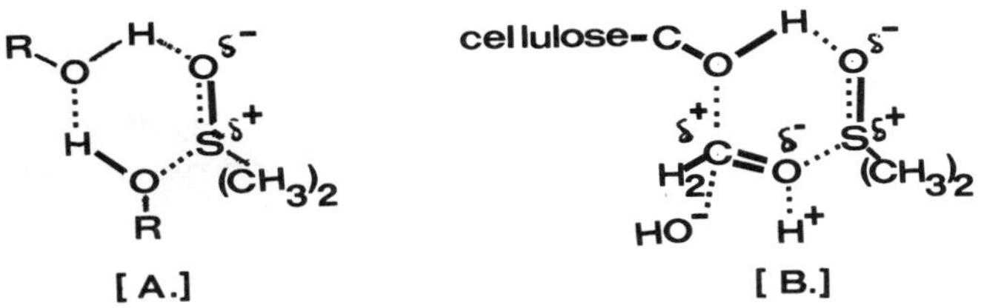

Figure 1. Proposed DMSO: Paraformaldehyde complexes of alcohols (A) and cellulose (B).

It is then likely that a comparable complex of DMSO and cellulose can exist.
And the existence of such a transient species would interfer with the intra-
and intermolecular hydrogen bonding of cellulose which is responsible for
its unusual physical, chemical and solubility properties. The interaction of
the complexes, paraformaldehyde, and traces of water are therefore believed to
result in the formation of cellulose methylols such as those shown in Figure
1B. Further, it is postulated that it is this complex of cellulose DMSO:PF
that is responsible for the solubilization of cellulose. One author (Johnson,
1976) has proposed that the similarities in solubility parameter observed when
one compares DMSO (12·0H), formaldehyde (12·6H), and cellulose (13.0 - 14.0H)
reflect the similar cohesive energy densities associated with these materials
and their favorable solution thermodynamics.

Knowing that solvents of similar solubility parameter (11-15H) do not precipitate
the solubilized cellulose in DMSO, we mixed polymers with solubility parameters
in this range, and formed clear homogeneous solutions. The solutions were maintained at all proportions of cellulose and polymer. We found that polyacrylonitrile) (PAN, 12.0H), polyvinylpyrrolidone (PVP, 12.5), and poly(vinyl
alcohol) (PVA, 12.0H) formed homogeneous cellulosic polyblend solutions.

Polymers with solubility parameters differing much from 11-14H were insoluble
and formed suspensions. These polymers included polystyrene (9.0H), poly(methyl-
methacrylate) (9.1H), poly(vinyl acetate) (9.2H), poly(ethylene terephthalate)
(Dacron, 9.5H), and poly(hexamethylene adipamide) (Nylon-66, 15.0H).

Cellulosic polyblends of PAN, PVP and PVA which were 1.0 percent cellulose and
0.3, 1.0 and 3.0 percent of the blended polymer were prepared. As shown in
Figure 2 the viscosities of these homogeneous solutions varied from 3,000 to
6,500 cps as a function of the relative amount of the blended polymer in
solution. The cellulose/PAN solutions were pale yellow clear solutions whose
viscosities were greatly effected by the composition of the solutions. The
cellulose/PVA solutions were water clear and exhibited a composition viscosity
relationship similar to that of the cellulose/PAN polyblend solutions. The

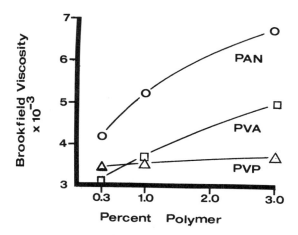

Figure 2. The Effect of Varying Blended Polymer Concentrations on the
           Viscosity of 1.0% Solutions of Cellulose in DMSO:PF.

cellulose/PVP solutions were also water clear. These solutions differed from the other two cellulosic polyblend solutions in that their viscosities were not as sensitive to changes in the cellulose/blended polymer ratio.

Cellulosic polyblend films were prepared by casting the polyblend solutions on glass plates as described above. The cellulose/PAN films were all tough transparent materials which were characterized by a faint yellow coloration. These films were easily handled and could be easily flexed without apparent fatigue or failure. The cellulose/PVP films were esthetically superior to the other cellulosic films. They were clear colorless films which varied greatly in flexibility as a function of their exposure to moisture. The cellulose/PVA films were similar.

The water absorption characteristics of these cellulosic films are summarized in Figure 3 which presents the average percent weight increase observed for films of varying composition. As shown in Figure 3 the cellulose/PVP films were most sensitive to water absorption. Both the PAN and the PVA films showed slight weight increases as the percent of the blended polymer was increased. These observations are in line with the hydrophilic nature of PVP.

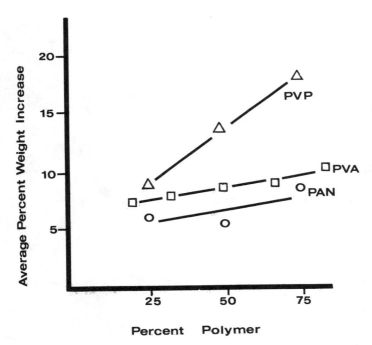

Figure 3:   Average percent weight increase observed when cellulose polyblend films are dried and then stored in a 62.5 percent relative humidity environment.

Average tensile strengths of the cellulose/PAN films are presented in Table II. The cellulose/PAN films were superior to the other films in tensile properties. As the percent PAN in the films was varied from 25 to 75 percent the tensile strength varied from 130 to a maximum of 544 kg/cm^2 for the 50 percent PAN cellulosic films. Both the tensile strength and the percent elongation reached maximum values at the 50 percent PAN level. The stress strain curves of these films indicated that they were hard – tough films.

TABLE II

Average Tensile Strength of Cellulose/PAN Films

Percent PAN	Tensile Strength, kg/cm^2	Elongation, %
75	357	4.8
50	544	5.5
25	130	4.1

SEM observations of the morphologies of the cellulosic films showed that the cellulose/PAN and the cellulose/PVP films were solid. The cellulose/PVA films were however, porous. The diameter of the pores observed in these films varied proportionately with the percent PVA in the films.

This can be noted by comparing the representative SEM micrographs: Figure 4 (cellulose-PVA, 80:20), Figure 5 (cellulose-PVA, 60:40), and Figure 6 (cellulose-PVA, 40:60). The average pore size varied from 5 to 80 microns as the percent PVA in the films was changed from 20 to 80 weight percent.

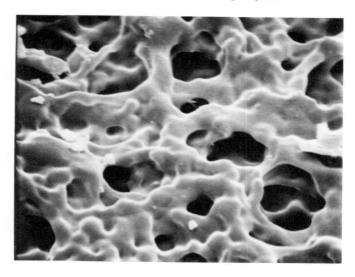

Figure 4.   Scanning Electron Micrograph of a Cellulose (80%)–Poly(Vinyl Alcohol) (20%) Film at 35,500x.

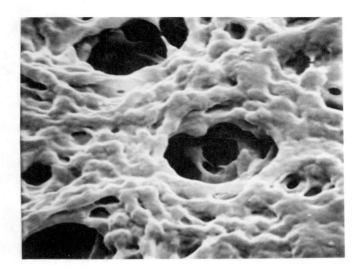

Figure 5.   Scanning Electron Micrograph of a Cellulose (60%)-Poly(Vinyl Alcohol)
            (40%) Film at 35,500x.

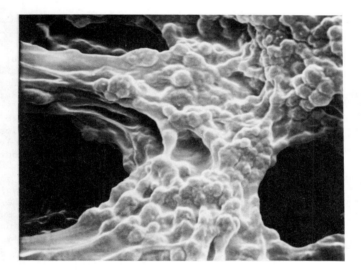

Figure 6.   Scanning Electron Micrograph of a Cellulose (40%)-Poly(Vinyl Alcohol)
            (60%) Film at 35,500x.

The relationship of average pore diameter and percent PVA is shown more clearly in Figure 7.

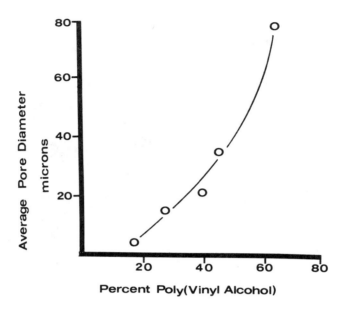

Figure 7.   Average Pore Diameter Observed in the SEM Inspection of Cellulose-
            Poly(Vinyl Alcohol) Polyblend Films at 71,000x.

The solubility of cellulose only in DMSO:PF mixtures, the miscibility of the stock solutions with basic and/or other similar polar aprotic solvents and the solubility of PAN, PVA and PVP in the DMSO:PF solutions of cellulose reflect the importance and functionality of solubility parameter considerations in the preparation and manipulation of DMSO:PF solutions of cellulose. Solvents and other polymers which were not within 2.0 Hildebrand units of the DMSO:PF solvent system (12.0 to 12.5) were miscible with the cellulose solutions.   It is thus apparent that the enthalpies of mixing of the solvents and polymers with DMSO:PF solutions of cellulose approaches zero (Seymour, 1971).

The film forming qualities and morphologies of the cellulosic films results from the homogeneous nature of the cellulose polyblend solutions and the relative solubilities of the cellulose and the blended polymers in the solvent system (DMSO:PF) as the DMSO and excess formaldehyde are vaporized during the curing process.   Both the tensile strength and the porosity of the films are related to the relative solubilities of the two polymers in the solvent and the evaporation of the solvent (Ott,Spurlin, and Grafflin, 1971; Sefton and Merrill, 1976).

The water absorption characteristics of the films is dictated by the relative hydrophilicities of PVA, PVP, PAN and cellulose. PVP is the most hydrophilic and as expected these films display the greatest water absorbtivity. PAN is the least hydrophilic and its cellulosic films display the lowest water absorbtivity. The plasticizing effect of water on the PVP and the PAN cellulosic films is the direct result of water absorption and increased molecular mobility.

CONCLUSION

It is concluded that DMSO:PF solutions of cellulose are miscible with polar aprotic solvents having solubility parameters between 10 and 14. Stock solutions of cellulose may also be blended with PAN, PVP and PVA to form homogeneous solutions which may be used to cast films at temperatures between 50° and 70°C. The hydrophilicity, tensile strength and porosity of these films may be varied over wide ranges as a function of the blended polymer selected and the relative amounts of cellulose and blended polymer present in the film.

ACKNOWLEDGEMENT

The authors wish to thank Mr. Robert Keith, Department of Electrical Engineering, and Mr. Samuel Young, Department of Chemical Engineering, for their assistance in obtaining the scanning electron micrographs and tensile strength measurements. Both are with the University of Houston. This work was supported in part by grants from the Robert A. Welch Foundation and ACS/ Petroleum Research Fund.

REFERENCES

Casu, B., M. Reggiani, G. G. Gallo, and A. Vigevani (1964). Tetrahedron Lett., 39, 2839.

Casu, B., M. Reggiani, G. G. Gallo, and A. Vigevani (1965). Tetrahedron Lett., 27, 2253.

Casu, B., M. Reggiani, G. G. Gallo, and A. Vegevani (1966). Tetrahedron Lett., 22, 3061.

Chapman, O. L. and R. W. King (1964a). J. Amer. Chem. Soc., 86, 1256.

Chapman, O. L. and R. W. King (1964b). J. Amer. Chem. Soc., 86, 4968.

Jacob, S. W., E. E. Rosenbaum, and D. C. Wood (1971). Dimethyl Sulfoxide, Vol. 1 Marcel Dekker, New York.

Johnson, D. C., M. D. Nicholson, and F. C. Haigh (1975). 8th Cellulose Conference, Syracuse, New York.

Johnson, E. L. (1976). The Synthesis and Characterization of Cellulose Esters In DMSO:PF Solutions. Doctoral Dissertation, University of Houston, Houston.

Ott, E., H. Spurlin, and M. W. Grafflin (1971). Cellulose and Cellulose Derivatives, Vol. V Interscience Publishers, Inc., New York. 644p.

Rader, C. P. (1966). J. Amer. Chem. Soc., 88, 1773.

Sefton, M. V. and E. W. Merrill (1976). J. Biomed. Mater. Res., 10, 33.

Seymour, R. B. (1971). Introduction to Polymer Chemistry, McGraw-Hill, New York. pp 44.

Seymour, R. B. and E. L. Johnson (1976). J. Applied Polymer Sci., 20, 3425.

CELLULOSE DISSOLUTION AND DERIVATIZATION IN LITHIUM CHLORIDE/N,N-DIMETHYLACETAMIDE SOLUTIONS

C.L. McCormick and T.S. Shen

Department of Polymer Science
University of Southern Mississippi
Hattiesburg, MS 39401

ABSTRACT

Dissolution of cellulose in N,N-dimethylacetamide/LiCl solvent is discussed. A mechanism involving association of a $Li[DMAc]^+$ macrocation with the chloride ion-cellulose complex is proposed based on NMR studies of model compounds and on polyelectrolyte behavior. Preparation of carbamate, ether, and ester derivatives of cellulose under homogeneous reaction conditions are reported.

KEYWORDS

Cellulose dissolution; homogeneous solutions; preparation of synthetic derivatives; polyelectrolyte behavior.

INTRODUCTION

In the past, the facile preparation of a wide range of derivatives of unmodified cellulose has been hampered by the lack of suitable organic solvents. Most synthetic reactions of cellulose, therefore, have been conducted heterogeneously, often resulting in much less uniform substitution than desired. In the absence of appropriate solvents, characterization of reactants is difficult and reaction yields are often low due to unfavorable kinetics.

Although a number of cellulose solvents have been discovered,[1-3] none has application to a wide range of organic reactions. Recently, we reported the synthesis of controlled release pesticides[4,5] and other derivatives[6] based on homogeneous reactions of cellulose in lithium chloride/N,N-dimethylacetamide (DMAc). In this paper, we report supportive data for a proposed mechanism of cellulose dissolution in DMAc/LiCl as well as preparation of ether, ester, and carbamate derivatives.

EXPERIMENTAL

Cellulose solutions were prepared by suspending 15.0 g of reagent grade cellulose (Baker) or cotton linters in 1500 ml of N,N-dimethylacetamide which contained

75.0 g of lithium chloride. The mixture was heated to 150°C and allowed to slowly cool to room temperature. If the cellulose had not completely dissolved, a second heating-cooling cycle was applied. The solution prepared from the Baker cellulose, referred to as solution "A," was used for all subsequent reactions ([η]=1.64 dl/g).

Carbamate derivatives of cellulose were prepared by adding dropwise a molar excess (more than three times the number of moles of anhydroglucose units) of the isocyanate in 50 ml of N,N-dimethyl acetamide to 100 ml of stirring solution A maintained at 90°C under nitrogen. Polymers were isolated by precipitation into a nonsolvent and then purified by Soxhlet extraction. Ester derivatives were prepared by reacting solution A with a molar excess (with respect to anhydroglucose units) of the appropriate acid chloride or anhydride. In some cases acid acceptors such as tertiary amines, ZnO, or BaO were utilized to improve yields. The cellulose derivatives were isolated in the same manner as previously described for the carbamate derivatives.

Cellulose ethers were synthesized by nucleophilic substitution reactions of the appropriate alkyl halides in the presence of base or by ring opening of ethylene oxide. Thus methyl cellulose, carboxymethyl cellulose, and benzyl cellulose were prepared by addition of a slight molar excess of methyl iodide, sodium chloroacetate, and benzyl chloride, respectively, to 100 ml of solution A in the presence of stoichiometric quantities of sodium hydroxide or tetraethyl ammonium hydroxide. All reactions were conducted at 50°C for 40 minutes except for the methylation reaction which was conducted in a programmed manner. Portions of methyl iodide were added at three forty minute intervals with temperature maintained at 50, 60, and 70°C during each addition. Hydroxyethyl cellulose was prepared by adding a molar excess of ethylene oxide to 100 ml of solution A maintained at 50°C. Approximately 1.5 ml of ethylene oxide was added five times at thirty minute intervals. Cellulose ether derivatives were isolated by precipitation into methanol or isobutanol followed by Soxhlet extraction and drying under reduced pressure at 50°C.

The degree of substitution (D.S.) for each derivative was determined by elemental analysis, infrared spectroscopy, or total hydrolysis. C^{13} and H^1 NMR studies were conducted (Jeol FX-90Q) on solutions of N,N-dimethylacetamide and model alcohols in the presence of LiCl and $ZnCl_2$ in attempts to elucidate the mechanism of dissolution. Cannon Ubbelohde four-bulb shear dilution viscometers were used to measure intrinsic viscosities of the cellulose solutions at different concentrations in the LiCl/DMAc solutions. Aging studies were conducted utilizing a Brookfield Viscometer.

RESULTS AND DISCUSSION

Lithium chloride in N,N-dimethylacetamide is a highly specific solvent for dissolution of cellulose. Under the conditions reported in the experimental section, an upper limit of 8% by weight cellulose was reached. Lithium chloride concentrations up to 11% (saturation) in DMAc were used. Interestingly, neither other lithium salts including bromide, iodide, nitrate, and sulfate nor other chloride salts such as sodium, potassium, barium, calcium, and zinc were effective. Additionally, N,N-dimethylformamide, despite its similarity in structure and solubility parameter, failed to dissolve cellulose when substituted for DMAc.

Although the mechanism of the dissolution process cannot be completely elucidated at present, rheological behavior of cellulose solutions and NMR analysis of model alcohols in DMAc/LiCl suggest possible interactions involving ion-pairing and hydrogen bonding. A model similar to that proposed by Panar and Beste[7] for dissolution of poly(1,4-benzamide) in a similar solvent system may be applicable.

The hydroxyl protons of cellulose may hydrogen bond to the chloride anion (in an analogous fashion to that proposed for the amide proton). The chloride ion appears to be associated with the $Li^+(DMAc)$ macrocation. The resulting charge-charge repulsions[3,9] or a bulking effect[10] would tend to "open" the polymer structure for further solvent penetration.

TABLE 1 Carbon[13] NMR Spectra of DMAc Solutions

Solute	Solvent	*CO	*CH3CO		N-*CH3
–	DMAc	169.80	21.40	34.62	37.70
ZnCl2 (4%)	DMAc	169.96	21.40	34.68	37.76
LiCl (6%)	DMAc	170.83	21.40	34.84	38.04

Downfield shifts of 1.0 ppm for the carbonyl C^{13} resonance and of 0.05 ppm in the N-methyl and acetyl proton resonances indicate strong complexation of the lithium cation with DMAc (Table I). This solvated cation exerts a profound influence on disruption of intermolecular hydrogen bonds of the cellulose molecules. No significant shift in the C^{13} carbonyl resonance was observed for $ZnCl_2$ in DMAc in a parallel experiment.

The specificity of the chloride anion with the $Li[DMAc]^+$ solvated cation suggests a more complicated mechanism than cation binding to the oxygen atom of the hydroxyl groups. Small downfield chemical shifts for hydroxyl protons of ethanol (Table II) and model compounds glucose and sucrose in the absence of DMAc are consistent with association with the chloride anion.

TABLE 2 PMR SPECTRA OF DMAC SOLUTIONS

Solvent	Ethanol			DMAC		
	CH3	CH2	OH	CH3C	N-CH3	
EtOH (6%); LiCl (6%)*	1.07	3.46	5.53	2.04	2.87	3.02
EtOH (ex); LiCl (6%)*	1.13	3.57	5.04	2.04	2.87	3.02
EtOH (4%); ZnCl2*	1.08	3.52	5.04	2.04	2.87	3.02
EtOH; LiCl (6%)	1.18	3.62	5.40			
EtOH (50%); DMAC (50%)	1.10	3.54	4.73	2.00	2.82	3.00
EtOH	1.16	3.57	5.32			
DMAC				1.98	2.80	2.98
LiCl (6%)*				2.05	2.87	3.03
ZnCl2 (4%)*				2.00	2.83	3.00

* in DMAC

Cellulose solutions in DMAc/LiCl show interesting aging behavior (Fig. 1) as followed by a decrease up to 48 hours for 1% solutions. After this initial period, however, a steady value was reached for the apparent viscosity. These aging effects are similar to those typically observed for water-soluble polymers and

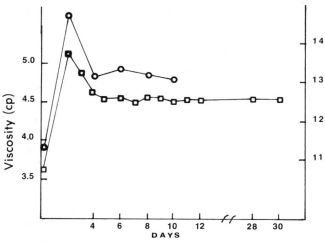

Fig. 1   Aging effects in DMAC/LiCl solutions

○ 1.4% cellulose, 6% LiCl
□ 1.0% cellulose, 7.8% LiCl

may be attributed to changing inter- and intramolecular hydrogen bonds.

Intrinsic viscosity was determined for a 5% LiCl/DMAc cellulose solution by plotting the reduced viscosity, η_{sp}/c vs concentration (Fig. 2).   Extrapolation of the linear Huggins plot to zero concentration yielded a value for intrinsic viscosity, [η], of 1.64 (correlation coefficient=0.994).   Interestingly, however, at lower lithium chloride concentrations (and low cellulose concentrations) deviations from linearity were observed suggesting possible polyelectrolyte

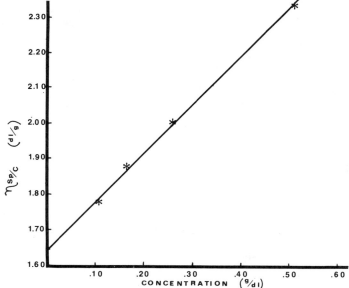

Fig. 2   Reduced viscosity vs concentration for a 5%
LiCl/DMAc solution of cellulose

behavior. At higher LiCl concentrations, ionic strength may be sufficient to overcome electrostatic interactions.

To further investigate possible polyelectrolyte behavior, intrinsic viscosity was measured as a function of ionic strength (Fig. 3). The linear correlation (0.992) and slope seem to indicate weak polyelectrolyte behavior consistent with the proposed model.

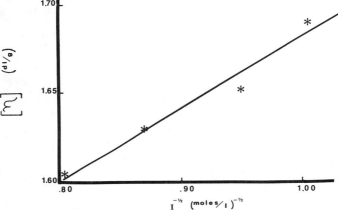

Fig. 3. Intrinsic viscosity of cellulose as a function of ionic strength of LiCl in DMAc

Cellulose solutions with concentrations of 0.015 to 0.30 g/dl and with LiCl contents varying 0.75 to 6.0% by weight of LiCl in DMAc showed Newtonian behavior in the 0.5 to 10 sec^{-1} shear rate region. These studies were conducted with Cannon Ubbelohde 4-bulb shear dilution viscometers. Attempts are presently underway to investigate a wider range of shear rates and polymer concentration utilizing a Haake^{R} rotational viscometer.

#	PPM
1	103.537
2	74.496
3,5	76.339
4	79.481
6	60.682

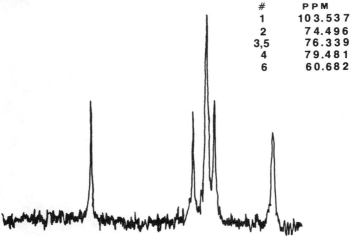

Fig. 4. C^{13} NMR spectrum of cellulose in 5% LiCl/DMAc

A significant feature of the DMAc/LiCl solvent is that characterization of the dissolved cellulose is possible by a number of direct instrumental techniques.

For example, the C^{13} NMR spectrum of cellulose is shown in Fig. 4. Light scattering and size exclusion chromatography studies are also underway in our laboratories at this time to determine molecular weight and size information.

The most important contribution of the LiCl/DMAc solvent for cellulose, however, is its applicability to a rather wide range of synthetic organic reactions. Not only can reactants and products be more thoroughly characterized, uniform substitutions should be possible since these are performed under homogeneous reaction conditions. Table 3 lists selected ether, carbamate, and ester deriva- tives of cellulose prepared to date. Moderate-to-high degrees of substitution have been obtained although no attempts have yet been made to optimize conditions (temperature, pressure, catalyst, etc.) for maximum yields.

TABLE 3   Cellulose Derivatives Prepared in LiCl/DMAc Solutions

R	Sample #	D.S.
⟨O⟩— N-C- (O, H)	1a	2.15
CH₃—⟨O⟩— N-C- (O, H)	2a	0.81
CH₃—⟨O⟩— S-N-C- (O, O, H)	3a	0.70
Cl—⟨O⟩— N-C- (O, H)	4a	0.12
(CH₃)₃C-C⟨ ⟩N-NH-C- ... C-SCH₃	5a	0.98
CH₃-C- (O)	2a	1.95
CH₃-C- (O)	2b	1.80
⟨O⟩-C- (O)	2c	1.54
CH₃CCl₂C- (O)	2d	0.50
Cl—⟨O⟩— O-CH₂-C- (Cl, O)	2e	0.66
CH₃-	3a	1.20
CH₃-	3b	0.80
⟨O⟩-CH₂-	3c	0.60
HOOC-CH₂-	6a	2.53
HOCH₂CH₂-	7a	1.80

CONCLUSIONS

The N,N-dimethylacetamide/lithium chloride solvent will allow dissolution of up to 8% cellulose by weight. A mechanism has been proposed in which hydroxyl protons on the cellulose hydrogen bond to the chloride ion which is in turn associated with the Li[DMAc]$^+$ macrocation complex. Carbon^{13} and proton NMR studies support the above proposal. Viscosity studies indicate solution stability after initial changes in hydrogen bonding.

Newtonian flow behavior was observed for the cellulose solutions over the range of shear rates and concentrations studied. Polyelectrolyte behavior is observed in a plot of intrinsic viscosity as a function of ionic strength of added LiCl salt in DMAc. A number of synthetic derivatives of cellulose have been prepared under homogeneous conditions in this solvent system. Additional studies into the mechanism of dissolution of cellulose are in progress including the role of associated water.

REFERENCES

Phillip, B., H. Schleicher, and W. Wagenknect, (1977). Chemtech, 702-709.
Turbak, A.F., R.B. Hammer, R.E. Davies, and H.L. Hergert, (1980). Chemtech, 51-57.
Hudson, S.M. and J.A. Cuculo, (1980). J. Macromol Sci-Rev. Macromol. Chem. C18(1), 1-82.
McCormick, C.L. and D.K. Lichatowich, (1979). J. Poly Sci:Polymer Letters, 17, 478-484.
McCormick, C.L., D.K. Lichatowich, J.A. Pelezo, and K.W. Anderson, (1980). Modification of Polymers, C.E. Carraher, Ed., ACS Symposium Series No. 121, 371-380.
McCormick, C.L., U.S. Patent # 4,278,790.
Panar, M. and L.F. Beste, (1977). Macromolecules, 10, 1401-1406.
Parker, A.J., (1962). Quarterly Rev. 16, 163.
Spurlin, H, (1955). Chapter H. In Spurlin, H. and M. Graffin (Interscience), Cellulose and Cellulose Derivatives, 4(3).
Gruenwald, E. and E. Price, (1964). J.Am.Chem.Soc. 86, 4517.

ETHYL CELLULOSE ANISOTROPIC MEMBRANES

Thomas C. Shen[*] and I. Cabasso[**]

Gulf South Research Institute, Polymer Dept. New Orleans

[*] Abbott Laboratories, North Chicago, Ill. 60064

[**] State University of New York, Chemistry Department

ABSTRACT

An ethyl cellulose anisotropic porous membrane with high flux rate, uniform sponge type support structure and pore size of less than 150 Å has been prepared. The effects of solvents, polymer concentration and casting conditions on membrane structure are discussed. Controlling the rate of polymer precipitation to form the optimal membrane structure is also presented.

KEY WORDS

Ethyl cellulose; anisotropic porous membrane; solubility parameter; ultrafiltration membranes.

INTRODUCTION

Synthetic membranes - microfiltration and ultrafiltration - have gained increasing prominence as a simple and convenient means for separation (concentrating, purifying, and fractionating solutions containing colloids or solutes of large molecular weights) (Flinn, 1970; Hwany, 1975). These membranes exhibit a microporous structure with high solute retention efficiencies and high solvent hydraulic conductivity rates.

There are several techniques in use for preparing porous membranes, e.g., the polysalt formation process, the phase inversion process, the particle-track etching process, and others (Michaels, 1961, 1965a, 1965b; Van Oss, 1970; Fleisher, 1963; Kesting, 1971). A major fraction of commercially available membranes are produced by the phase inversion process, whereby a polymer solution is coagulated in a nonsolvent bath to yield an anisotropic (asymmetric) structure. This type of structure consists of a very thin and relatively dense skin, supported by a porous substructure (as shown in Fig. 1). Perfection of this technique yields a highly reproducible membrane with filtration properties that can be controlled for each specific application by the proper choice of polymer, solvent, precipitant, and by manipulation of other preparation parameters. The mechanism of membrane formation (Loeb, 1962; Sourirajan, 1970; Frommer, 1970, 1972; Strathmann, 1971a, 1971b;

Saier, 1974; Frommer, 1973; Guillotin, 1977; Strathmann, 1977; Koenhen, 1977) by
the phase inversion process is quite complex, and involves myriad parameters - the
contribution of which is not yet well defined.

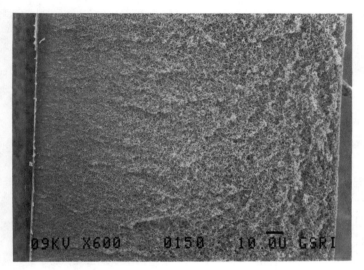

Fig. 1. Scanning electron microscope (SEM) photograph of
asymmetric ethyl cellulose porous membrane prepared
from solvent system 1 at 20% polymer concentration.

EXPERIMENTAL

Membrane Preparation
Ethyl cellulose porous membranes were prepared from three grades of polymer:
Ethocel[R] Standard 20, Ethocel[R] Medium 50, and Ethocel[R] Medium 100, which are manu-
factured by the Dow Chemical Company. The degree of ethoxy substitution ranged
from 2.25-2.58.

Ethyl cellulose solutions were prepared, filtered and cast in a dust-free laminar
flow hood. The solutions were cast on glass at clearances of 6-11 mils with evaporation periods ranging from 5-180 sec before coagulation in an aqueous bath
(Table 1).

Testing and Evaluation
All the membranes were tested under hydrostatic pressures ranging from 50-800 psi.
The detailed procedure, cell structure and testing system have been previously
reported (Shen, 1981).

Membrane Morphology Studies
Scanning electron micrographs were taken by a Jeol JSM-35C instrument. All specimens were gold-palladium coated. The membranes were fragmented in liquid nitrogen
to obtain the cross-section morphologies.

RESULTS AND DISCUSSION

Solvent Selection
The three-dimensional solubility parameter system (Broens, 1977) has been used for

many years in interpreting the solvent-polymer interaction.   This concept has also been adopted in selecting the solvents for the preparation of asymmetric membranes (Hansen, 1971).   In the early 1970's, Klein and Smith (1972) reported a set of working rules for use of the solubility parameters associated with the anisotropic coagulation of a cast layer of polymer solution in a nonsolvent (thus yielding an "asymmetric" membrane).

1) The casting composition solubility parameter (δ) should be near the solubility boundary facing the quench medium.
2) A volatile solvent component should be such that its loss will move the composition out of the solubility area, rather than into it.
3) The solids content at the solution boundary must be high in order to cause a rapid phase transition from solution to gel.
4) All components of the solvent system should be miscible with the quench medium.

In order to apply these rules in preparing ethyl cellulose anisotropic porous membranes, the solubility diagram of ethyl cellulose was established (Fig. 2).   Several solvent systems were selected and formulated.   The results of testing flux and salt rejection under applied pressures (Fig. 3) could not be correlated with solubility parameters of the solvent system, as explained in the following section.

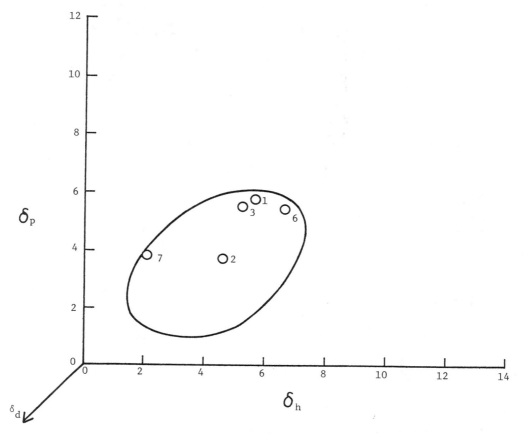

Fig. 2.   Solubility diagram for ethycellulose (degree of substitution 2.25, 100 cps). (solvent system listed in Table 1)

TABLE 1. ETHYL CELLULOSE MEMBRANES (CASTING CONDITIONS AND PERFORMANCE)

Membrane Number	Ethyl Cellulose (wt%)	Methanol (g)	Dioxane (g)	Methyl Acetate (g)	Isopropanol (g)	Formamide (g)	Solvent Density (g/cc)	Salt Rejection (%)	Flux (gfd)
1	11.03	21	57	--	--	15	0.922	0	800
2	11.60	--	60	--	30	--	0.944	17	38
3	11.43	--	90	--	45	20	0.963	0	70
4	17	18	102	--	--	19	1.02	79.94	0.09
5	17	32.8	95	--	--	15	1.006	82.52	0.22
6	12.0	24	--	60	--	16	0.945	0	0
7	12.0	--	88	--	--	--	1.033	96	0.4

Membranes 1-3 were cast at 11 mils with an evaporation period of 40 sec, and quenched at 22°C (polymer viscosity = 100 cps). Membranes 4-5 were cast under the same conditions, with the exception of the evaporation period, which was 10 sec, and the viscosity, which was 50 cps. All membranes were tested with 2000 ppm (NaCl) aqueous solution at 100 psi.

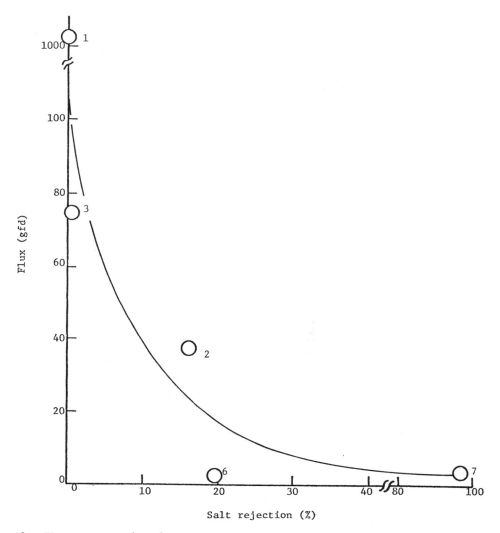

Fig. 3.   Flux vs. salt (NaCl) rejection of ethyl cellulose membranes at 100 psi.

Solvent-Nonsolvent Exchange
Several types of sublayers were observed by scanning electron microscopy. Among
these structures, two predominant morphologies - finger-like and sponge-like - are
clearly observed.   The experimental results indicate that the way these layers form
depends on the rate of precipitation of the polymer (Guillotin, 1977; Klein, 1972).
The rate of precipitation of the polymer is determined by the rates of solvent
diffusing out of, as well as the rates of nonsolvent diffusing into, the polymer at
the interface (Strathmann, 1971b; Sanderson, 1978).   Several factors affect the rate
of precipitation of the polymer, such as solvation power, polymer concentration,
and solvent/nonsolvent interaction.

   Solvation power.   Solvation power as related to the polymer and solvent interaction

is also related to the solubility parameters of the two. A higher solvation power of a solvent means the polymer uncoil in the solution, and needs a
larger amount of nonsolvent to precipitate the polymer. It can also be predicted
by the difference of the solubility parameter.

$$\Delta\delta = \sum_{i=1}^{3} (\delta_{ip}^2 - \delta_{is}^2) \tag{1}$$

As $\Delta\delta$ decreases, the solvation power of the solvent increases. In general, the
lower the solvation power, the higher the rate of precipitation of the polymer
becomes, which may produce a finger-like support structure, even when all other
factors tend to produce a sponge-like structure, as shown in several reports
(Frommer, 1973; Kesting, 1977; Cohen, 1979).

Solvent-nonsolvent interaction. The general effect of the membrane casting solution on the properties of the membrane is shown in Fig. 4, a three-component
phase diagram. Here, the polymer is ethyl cellulose and the nonsolvent is water
(solvents are listed in Table 1); OP is a boundary layer. The region on the
right side of the boundary is a two-phase area, the upper left side represents a
gel formation area, and the lower left side indicates a homogeneous solution
area. The path AA'O indicates that the cast solution evaporated to form a dense
film; point A indicates that the solvent system contains a small amount of nonsolvent, e.g., formamide. Without nonsolvent present, point A would be located on
the nonsolvent baseline.

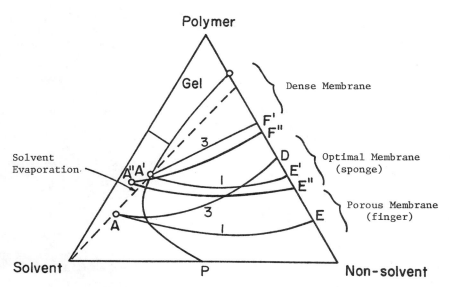

Fig. 4.   Phase diagram of the ethyl cellulose-solvent-water system, showing the
          precipitation paths at various rates of solvent-nonsolvent exchange.

Path AE represents a membrane formed during the precipitation step – water enters
the cast film faster than the solvent diffuses out. This forms a highly porous
membrane having the finger-like substructure (as indicated in Fig. 5, and in
Table 1 by solvent system 1 composition). Path AD indicates that the solvent

diffuses out faster than water diffuses in, to form a membrane with small pores and a sponge-like substructure. This applies to solvent systems 3, 6 and 7, as shown in Table 1. Solvent system 7 followed path AD and formed a dense film. This was confirmed by Frommer, and colleagues (1973) who measured water influx and solvent outflux of a 20% cellulose acetate solution, and showed that the solvent, dioxane, exits faster than the nonsolvent, water, enters. In order to make the solvent and nonsolvent exchange rates equivalent, formamide was added. The resulting membranes are shown in Fig. 6. The finger-like structure in the support layer indicates that the water entering the cast layer increases as formamide is added (i.e., precipitation rate increases). A cosolvent, methanol, was added to control viscosity and to equalize the solvent and nonsolvent exchange rates. At higher polymer concentrations, the membrane cast from these three solvents has a macrovoid-free, sponge-like substructure (Fig. 1).

Polymer concentration. Path A"E" is followed for higher polymer concentrations of solvent system 1, while path AE is followed for low polymer concentrations. This is due to the viscosity of the solution, which slows the rate of nonsolvent diffusing in, thus making the solvent and nonsolvent rates almost equal. This forms a membrane with a sponge-like morphology. Fig. 5 demonstrates that as polymer concentration increases, the occurrance and size of the finger-like macrovoids decreases. If solvent system 7 is used, the path A"F" will be followed to form a very dense film, owing to the very low rate of water entering the cast film.

Evaporation time. In solvent system 1, the solvent evaporates from the top layer of the cast film and the polymer concentration increases at the skin layer, as indicated in Fig. 4, point A'. Since the exchange rate of nonsolvent decreases at this point, the rate of polymer precipitation also decreases and forms a sponge-like structure, following path A'E'. If solvent system 7 is used, the path A'F' may be followed to form a very dense membrane. Membrane 44-1 was allowed a 5 second evaporation period, and membrane 44-3, a 25 second evaporation period. The flux rate (Table 2) decreases from 75 gfd to 4.0 gfd, indicating the formation of a dense film (Fig. 7). A thicker skin layer and a denser sponge layer were formed with an evaporation period of 5 seconds.

Quench medium. The quench medium is a nonsolvent for the polymer (water was used here). The addition of salt or glucose slows the exchange of solvent and nonsolvent (thus slowing the rate of polymer precipitation), due to the reduction of chemical potential between the solvent and nonsolvent. In other words, path A"F" would be followed instead of path A"E" to form a denser membrane, as shown by Table 2 - low flux rates of membranes 44-4 and 44-5.

Fig. 8 summarizes the relationship of the membrane support structure to the rate of solvent-nonsolvent exchange. These five types of support structure are obtained from most polymers.

        CONCLUSIONS

The optimal casting formulation and conditions for preparing the ethyl cellulose anisotropic porous membrane have been defined. The resulting membranes possess a high flux (> 100 gfd/100 psig), are macrovoid-free with a uniform pore size for both skin and bottom surfaces.

A good hollow fiber was obtained using these casting conditions. The results of testing for molecular weight cut-off as well as the preparation of composite membranes will be published.

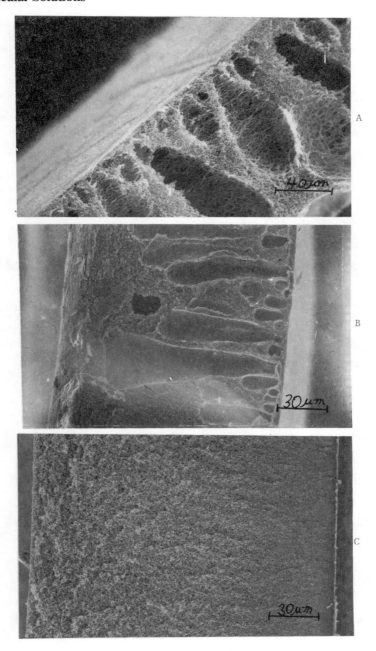

Fig. 5. Support layer morphology varies with polymer concentration. A) 13 wt% polymer   B) 15 wt% polymer   C) 20 wt% polymer.

Fig. 6.   Ethyl cellulose membrane prepared from A) dioxane, and B) dioxane and
          formamide solution.

TABLE 2.   CASTING CONDITIONS

Membrane Number	Evaporation Time (sec)	Thickness (mils)	Quenching Medium	Flux Rate At 50 psi (gfd)
44-1	5	5	22°C, water	75
44-2	5	8	22°C, water	80
44-7	5	11	22°C, water	85
44-6	5	5	6°C, water	78
44-4	5	5	22°C, 5% NaCl	8.4
44-5	5	5	6°C, 5% NaCl	6.2
44-3	25	5	22°C, water	4.0
44-8	45	11	22°C, water	0.1
2040*	80	11	22°C, water	160
2041*	120	11	22°C, water	116
2042*	80	11	4°C, water	120
2043*	120	11	7°C, water	120

* 12% ethyl cellulose

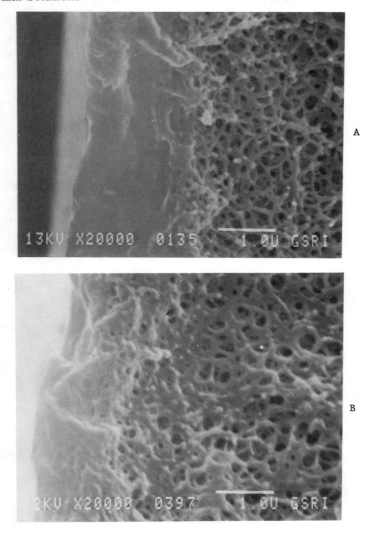

Fig. 7. Skin thickness at different evaporation times. A) 40 second evaporation time B) 5 second evaporation time.

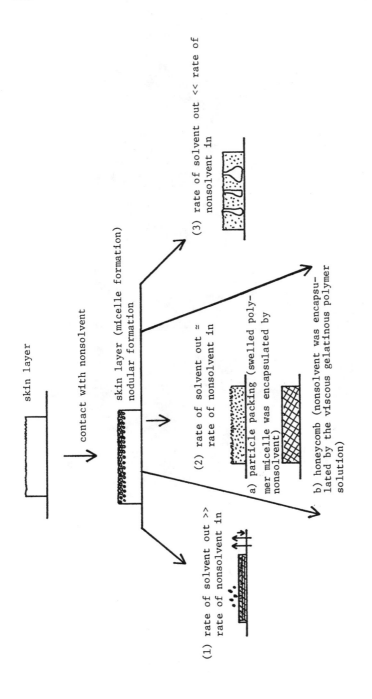

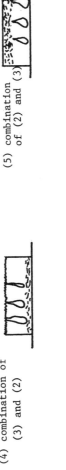

Fig. 8. Mechanism of membrane substructure formation.

ACKNOWLEDGEMENTS

We thank the U.S. Department of the Interior, Office of Water Research and Technology for supporting this study under contract 14-34-0001-0523, and Ms. B. Zimny
and Mr. R. LeBoeuf for their testing of the membrane.

REFERENCES

Broens, L, D.M. Koenhen, C.A. Smolders, (1977). Desalination 22, 205.
Cohen, Clande, (1979). J. Polym. Sci. 17 (Polymer Phys. Ed.), 477.
Fleisher, R., P. Price, and R. Walker (1963). Rev. Sci. Instrum 34, 510.
Flinn, J.E., Ed., (1970) Membrane Science and Technology. Plenum Press, New York.
Frommer, M.A., I. Feiner, O. Kedem, and R. Bloch (1970). Desalination 7 , 393.
Frommer, M.A., and D. Lancet (1972) in Reverse Osmosis Membrane Research, H.K.
    Lonsdale and H.E. Podall (Ed.). Plenum Press, New York.
Frommer, M.A., and R.M. Messalen (1973). Ind. Eng. Chem. Prod. Res. Dev. 12, 328.
Guillotin, M., C. Lemoyne, C. Noel, and L. Monnerie (1977). Desalination 21, 165.
Hansen, C.M. and A. Beerbower (1971). Solubility Parameters, in Encyclopedia of
    Chemical Technology, Supp. Vol., Wiley-Intersciences, New York.
Hwany, S.T. and K. Kammermeyer (1975) in Membranes in Separations, Wiley-Intersci-
    ence, New York.
Kesting, R.E. (1977) in Reverse Osmosis and Synthetic Membranes, S. Sourirajan (Ed.).
    National Research Council, Ottawa, Canada.
Kesting, R.E. (1971) in Synthetic Polymeric Membranes. McGraw-Hill, New York.
Klein, E. and J.K. Smith, (1972). I & EC Prod. Res. and Dev. 11 207.
Koenhen, D.M., et al., (1977) J. Appl. Polym. Sci. 21, 199.
Loeb, S. and S. Sourirajan (1962). Adva. Chem. Ser. 38, 117.
Michaels, A. (1965a). Ind. Eng. Chem. 57 (10) 32.
Michaels, A. H. Bixler, R. Hansslein and S. Fleming, (1965b) OSW Res. Develop. Prog.
    Rep. No. 194, U.S. Dept. of the Interior, Washington, D.C.
Michaels, A. and R. Miecka (1961). J. Phys. Chem. 65, 1765.
Saier, H.D., H. Strathmann, and U.V. Mylius (1974). Angew. Makr. Chem., 40/41, 391.
Sanderson, R.D. and H.S. Pienaar (1978), Desalination 25, 281.
Shen, Thomas C. and I. Cabasso (1981) Report to U.S. Dept. of the Interior, Office
    of Water Research and Technology, Contract No. 14-34-0001-0523.
Sourirajan, S, (1970) in Reverse Osmosis. Acad. Press, New York.
Strathmann, H., P. Scheible and R.W. Baker (1971b). J. Appl Polym. Sci 15, 811.
Strathmann, H., K. Kock, et al., (1975) Desalination 16, 197, (1977) 21, 241.
Strathmann, H., and P. Scheible (1971a). Koll. Z.U.Z. Polymer 246, 669.
Van Oss, C.J. and P.B. Bronson (1970). Sep. Sci. 5 , (1) 63.

WETTABILITY OF POLYMERS AND HYDROGELS AS DETERMINED BY WILHELMY PLATE TECHNIQUE

D.E. Gregonis*+, R. Hsu+, D.E. Buerger*, L.M. Smith*, and J.D. Andrade*+†

*Department of Materials Science and Engineering,
+Department of Pharmaceutics, and
†Department of Bioengineering, University
of Utah, Salt Lake City, Utah 84112

ABSTRACT

Polymers and hydrogel surfaces are investigated by contact angle procedures. A comparison between underwater captive air bubble contact angles and advancing and receding water contact angles using the Wilhelmy plate technique are studied using a sequence of hydrophobic to hydrophilic model polymers. The underwater captive bubble contact angle correlates most closely with the water Wilhelmy plate receding contact angle in most of the polymer systems. Contact angle hysteresis is found in all but the most wettable of the polymers but no exact trends are found. Alkyl derivatized agarose surfaces are also studied by the Wilhelmy plate procedure; although these surfaces exhibit strong protein interactions, little change in advancing and receding contact angle is observed with increasing degree of alkyl group derivatization.

KEYWORDS

Wilhelmy plate; contact angle; hydrophilicity; contact angle hysteresis; surface wettability; alkyl agarose surfaces; polymer surfaces.

INTRODUCTION

The study of the interface between an aqueous solution and a polymer surface is of considerable interest for the investigation of biological interactions. Contact angle methods are one of the few techniques capable of measuring the polymer-water interface directly. Contact angles are determined primarily by the outermost exposed atoms, possibly the outer 10 Å of a surface (Johnson, 1969). The hydrogel-water interface, however, is more diffuse than that of a hydrophobic polymer interface and the transition region between the bulk gel and free water may be on the order of 100 Å or greater in thickness. Our previous contact angle studies have utilized the underwater air and octane captive bubble technique in order to analyze the fully hydrated polymer surface (Andrade, 1979a, 1979b). Several drawbacks of this technique include the length of time for measurements and the difficulty to obtain reproducible angles on very hydrophilic surfaces. In addition, the captive bubble measurement provides only the static contact angle, and thus dynamic contact angles, which may offer further information about the interface, are not available.

The Wilhelmy plate technique (Wilhelmy, 1863) eliminates many of these difficulties.

Advancing and receding dynamic and static angles are easily obtained by this technique and are recorded on chart paper for permanent records. The first part of this Wilhelmy plate study correlates angles obtained with this technique and that obtained from underwater captive air bubble procedures. Poly(hydroxyethyl methacrylate) [PHEMA] and poly(methyl methacrylate) [PMMA] copolymers are used for this study. The second part of the study investigates advancing and receding angles obtained in a sequence of hydrophilic-hydrophobic triblock copolymers. These polymers were prepared by hydroboration of the butadiene block of styrene-butadiene-styrene (S-B-S) radial triblock systems. The last series of materials that were investigated consists of alkyl derivatized agarose surfaces. The alkyl agaroses bind quite strongly some proteins from an aqueous environment depending upon the size of alkyl group and the degree of group substitution and the hydrophobic character of the protein. These materials in bead form are commonly used in hydrophobic chromatography separation technique (Shatiel, 1974).

MATERIALS AND METHODS

The Wilhelmy plate apparatus incorporates a Scotts SRE500 mechanical testing machine and is used to raise and lower a beaker of 2X distilled water at a controlled speed of approximately 40 mm per minute. Situated above the beaker is a Cahn model RM-2 electrobalance which supports the test sample on a small thread. The balance is mounted separately and is vibrationally isolated. The mechanical tester and balance are contained in an insulated enclosure maintained at constant temperature (20°C) and humidity (30% RH). Electrical signals from the balance and cross head are fed to a X-Y recorder to obtain the wetting traces. Calibration of the balance is performed by the addition of a 200 mg tare weight to the sample. Other studies describe the measurement and equipment in more detail (Johnson, 1969; Smith, 1981). The captive bubble technique as performed in this work is also described elsewhere (Andrade, 1979; King, 1981).

The use of pure water in these experiments, especially when hydrating the samples for long periods of time, is very important. Water used for the wetting measurements and hydration studies found in this report is first deionized by passing it through a mixed ion exchange resin bed (Continental Water) and then distilled from a Barnsted Model YD 302 pyrogen removing still. The water is distilled a second time from an all glass system in which a small amount of potassium permanganate is added to degrade any organic substituents. The water is tested and shown to be free of pyrogens by the limulus lysate test and absent of bacteria. To this water was then added sodium azide (200 mg/1) and chlorox (30 µl/1, 1.5 ppm Cl final concentration) to prevent micro-organism recontamination. The surface tension of water is measured to be 72.6 ± 0.2 dynes/cm at 20°C using completely wetting glass microscope coverslip which was cleaned in chromic acid followed by a two minute helium radiofrequency glow discharge treatment at 200 µm Hg at 30 watts. In equation 1, θ is set equal to zero, and γ, the surface tension of water is thus calculated. For polymer hydration studies, chromic acid cleaned, all glass Coplin jars are used to store the samples.

The polymers to be studied are dip cast onto 24 x 50 mm microscope coverslips which are washed and cleaned to remove particulates and organic contaminants. The coatings are in the 1-3 micron range in thickness. Optical microscopy is used to verify surface uniformity. Polymer solutions (3% wt/vol) are filtered through 0.2 µ Fluoropore filters (Millipore Corp.) and stored in particulate free brown glass bottles. To improve polymer adhesion to the glass, silane treatments are sometimes required. A vapor phase silanization treatment, similar to that described by Haller (1978) is used. γ-Aminopropyl triethoxysilane (Aldrich Chemical) is used for the HEMA-MMA copolymer coatings and n-pentyl triethoxysilane (Petrarch Chemical) is used for the modified S-B-S triblock polymers. Agarose surfaces are deposited onto silanized clean glass coverslips. Copolymerization data for the HEMA-

MMA material is shown in Table I. Purified HEMA was donated by Hydro-Med Sciences and MMA was obtained from Aldrich Chemical Company. Copolymerization of HEMA and MMA has been shown to produce a random copolymer with a slight tendency for alternating addition (Patel, 1981; Okano, 1976).

TABLE 1   Hydroxyethyl Methacrylate (HEMA) - Methyl Methacrylate (MMA) Copolymerization Data

Copolymers (mole ratio)	Polymerization Solvent	Precipitation Solvent	Degree of Conversion
100% MMA	toluene	60-90 pet ether	34
99% MMA-1% HEMA	toluene	60-90 pet ether	44
95% MMA-5% HEMA	90% toluene - 10% ETOH	60-90 pet ether	48
75% MMA-25% HEMA	50% THF - 50% ETOH	H_2O	46
50% MMA-50% HEMA	50% THF - 50% ETOH	H_2O	44
25% MMA-75% HEMA	50% THF - 50% ETOH	H_2O	23
5% MMA-95% HEMA	ETOH	Diethyl ether	83
1% MMA-99% HEMA	ETOH	Diethyl ether	85
100% HEMA	MeOH	Diethyl ether	--

The Solprene radial S-B-S triblock copolymers were donated by Phillips Chemical Company. The styrene and butadiene homopolymers were obtained from Aldrich Chemical Company. Syndiotactic 1,2-butadiene polymer was obtained as a gift from Uniroyal Chemical Company. Characterization data of these materials is shown on Table 2. Proton nuclear magnetic resonance characterization to determine the amount and configuration of styrene and butadiene is measured on a Varian SC-300, 300 MHz instrument. Molecular weights are measured by gel permeation chromatography using μ styragel columns (Waters Associates) with nominal pore sizes of 5×10^2, 10^3. 10^4 and 10^5 Å. Tetrahydrofuran is used as the eluent and the column set was calibrated using narrow MWD polystyrene standards (Pressure Chemical). The butadiene segment of the S-B-S triblock copolymers is selectively hydroxylated using the hydroboration reagent, 9-borobicyclo [3.3.1] nonane, (9-BBN), in tetrahydrofuran. This borane is used to prevent crosslinking during the reaction which occurs when multi-reactive boranes are used (Levesque, 1971). Proton nuclear magnetic resonance shows the reaction to proceed in a quantative fashion. The polybutadiene double bonds in these polymers exist in both a cis and trans geometry with some 1,2 butadiene segments (Table 2). The hydroboration addition is nonspecific for the addition across the cis or trans double bond, but adds in an anti-Markovnikov manner (Brown, 1959) (Fig. 1) to the pendant 1,2-polybutadiene segments to produce primary hydroxyl groups. It is interesting to note that the equilibrium water swelling of the hydroxylated butadiene homopolymers is greatest for the 1,2 butadiene polymer, intermediate in the cis-trans polybutadiene and lowest in the cis-polybutadiene (Table 3) and may suggest a difference in hydroboration addition between the pure cis and cis-trans polybutadiene. Also shown in Table 3 are the water contact angles obtained from 24 hour water equilibrated samples. The water content of the polymer is calculated from the weight of water in the hydrated sample multiplied by one hundred (Gregonis, 1976).

The polysaccharide agarose is a naturally occurring polymer isolated from red seaweed. It consists of a repeating 1,3 linked β-D-galactopyranose and a 1,4 linked 3,6 anhydro-α-L-galactopyranose structure as shown in Fig. 2. This polysaccharide aggregates upon cooling from an aqueous solution and produces a high water content gel with surprisingly strong mechanical properties. The agarose used in this work is obtained as uncrosslinked Sepharose 4B-200 beads (Sigma Chemical Co).

TABLE 2   Styrene-Butadiene-Styrene (S-B-S) Triblock Copolymer Analysis

Copolymer	Wt% Styrene	Butadiene Configuration %			$M_w \times 10^{-5}$	$M_n \times 10^{-5}$
		Cis	Trans	1,2-		
styrene	100	--	--	--	8.72	1.83
solprene K	77	39.7	45.5	4.8	2.31	0.23
solprene 481	48	41.7	53.1	5.2	7.23	2.64
solprene 414	40	40.8	47.3	11.9	1.70	1.39
solprene 416	30	38.4	54.5	6.1	4.79	1.15
solprene 422	20	40.5	46.5	13.0	4.43	1.65
cis, trans-butadiene	0	39.3	51.4	9.3	6.87	1.36
cis-butadiene	0	99	--	1	7.15	1.45
syndio-1,2-butadiene	0	--	--	93	1.14	0.46

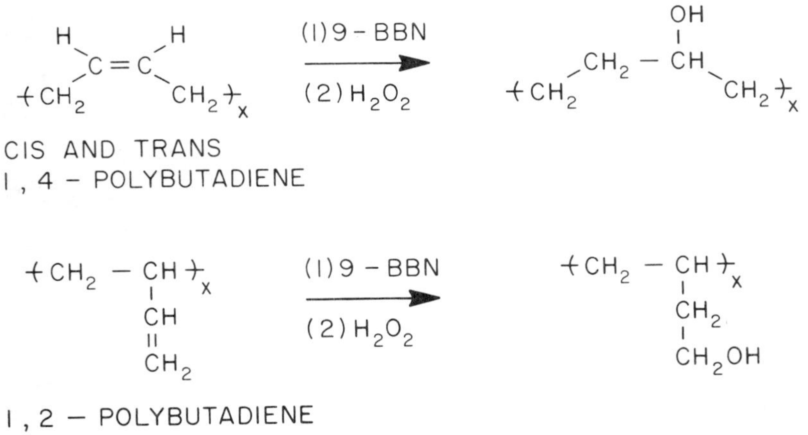

CIS AND TRANS
1,4 - POLYBUTADIENE

1,2 - POLYBUTADIENE

Fig. 1   Hydroxylation of olefin containing polymers by hydroboration procedures (9-BBN = 9-borobicyclo [3.3.1] nonane).

Agarose has been selectively derivatized to provide hydrophobic alkyl groups co-valently linked to the polysaccharide and is used as a selective protein separa-tion technique called hydrophobic chromatography (Svinivasen, 1980; Ochoa, 1978). Three general methods are commonly used for this derivatization procedure (Hjer-ten, 1974; Bethell, 1979; Cuatrecasas; 1970) but all of these procedures use or produce covalently crosslinked gels. Thus, they may be used in the configuration in which they were derivatized, but are not able to be redissolved for coating surfaces or devices. We have developed an alternate derivatization procedure which overcomes this disadvantage. In addition, our procedure provides easily prepared radioisotopically labeled alkyl groups for measurement of the degree of derivatization. This derivatization procedure is shown in Fig. 3

TABLE 3   Hydroborated Styrene-Butadiene-Styrene (SBS) Copolymer Data

Copolymer	Wt% Styrene	% Equil. Water	Captive Air Bubble θ	Advancing θ	Receding θ	Hysteresis advθ-recθ
styrene	100	< 1	87.8±1.3	93.8±2.8	67.1± 2.9	26.7
solprene K	77	3.3±0.4	34.3±2.3	81.6±7.4	16.6± 4.6	65.0
solprene 481	48	12.1±0.4	24.8±1.5	77.4±0.8	19.7± 1.8	57.7
solprene 414	40	12.4±0.3	-------	80.5±1.2	11.8±10.2	68.7
solprene 416	30	14.0±0.6	20.4±3.4	79.3±0.8	22.3± 0.7	57.0
solprene 422	20	14.6±0.6	22.9±1.6	78.3±0.8	19.9± 2.4	58.4
cis, trans-butadiene	0	17.5±0.4	23.4±3.6	87.9±2.1	16.7± 1.4	71.2
cis-butadiene	0	11.7±0.6	25.3±2.4	86.3±1.7	22.4± 0.8	63.6
syndio, 1,2-butadiene	0	19.6±0.3	23.5±4.4	83.3±1.4	17.8± 0.5	65.5

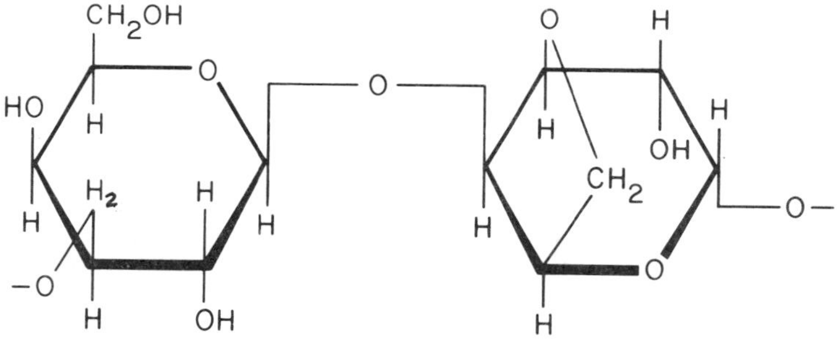

Fig. 2.   Repeating sub-unit of agarose.

The key development of this reaction procedure was finding the aqueous agarose beads can be exchanged with an aprotic solvent (acetone) without changing the agarose structure. After derivatization, the beads are then re-exchanged with water for use as hydrophobic column supports or dried and then redissolved in dimethyl sulfoxide for use in solution coatings. The agarose loses its ability to redissolve in water after low amounts of alkyl group derivatization. The alkyl group-agarose bond is determined to be stable in a distilled water environment for over three months. The agarose in this study was derivatized with varying amounts of n-butyl or n-dodecyl groups. The degree of derivatization is determined by liquid scintillation procedures. The agarose beads are first digested in hydrogen peroxide-perchloric acid solution and then counted in Biofluor cocktail (New England Nuclear). The degree of derivatization is reported as μmoles alkyl group per ml packed beads. A value of 15 μmoles alkyl group per ml packed gel corresponds approximately to 0.1 moles alkyl residues per mole anhydrodisaccharide repeat unit.

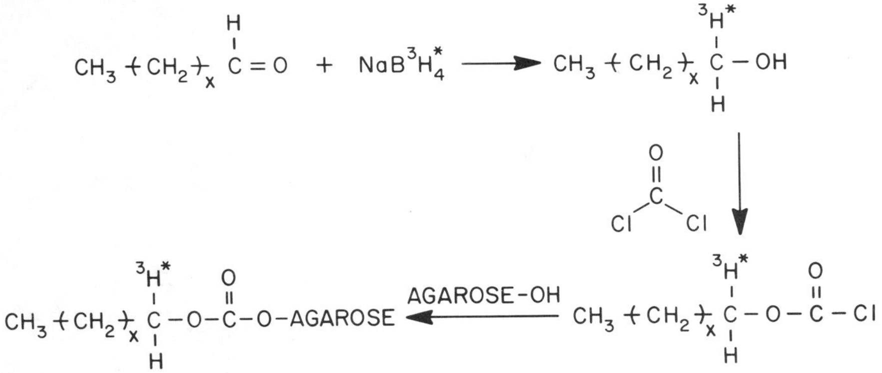

$$CH_3 + CH_2 \xrightarrow{}_X \overset{\overset{\displaystyle H}{|}}{C} = O \quad + \quad NaB^3H_4^* \longrightarrow CH_3 + CH_2 \xrightarrow{}_X \overset{\overset{\displaystyle {}^3H^*}{|}}{\underset{\underset{\displaystyle H}{|}}{C}} - OH$$

$$\overset{O}{\overset{||}{\underset{Cl\diagup^{\displaystyle C}\diagdown Cl}{C}}} \Big\downarrow$$

$$CH_3 + CH_2 \xrightarrow{}_X \overset{\overset{\displaystyle {}^3H^*}{|}}{\underset{\underset{\displaystyle H}{|}}{C}} - O - \overset{\overset{\displaystyle O}{||}}{C} - O - AGAROSE \xleftarrow{AGAROSE-OH} CH_3 + CH_2 \xrightarrow{}_X \overset{\overset{\displaystyle {}^3H^*}{|}}{\underset{\underset{\displaystyle H}{|}}{C}} - O - \overset{\overset{\displaystyle O}{||}}{C} - Cl$$

Fig. 3. Preparation and covalent bonding of α-tritiated alcohols to agarose.

DISCUSSION AND RESULTS

Contact angles may be measured by several techniques (Johnson, 1969). In this study a comparison between the underwater captive air bubble contact angle (Fig. 4) and the Wilhelmy plate method (Fig. 5) is investigated using the HEMA-MMA co-polymers. In general, the angle which the water makes with the surface can be measured directly with both the captive bubble and the Wilhelmy plate technique. This is done with the captive bubble contact angle method, using a horizontal microscope and measuring drop dimensions (King, 1981), but with the Wilhelmy plate apparatus it is most convenient to measure the force on the slide as it is immersed and withdrawn from the water. The advancing and receding angles are calculated from straight line extrapolations of the advancing and receding buoyancy slopes at zero depth of immersion (Smith, 1981). By this procedure the buoyancy factor in Eq. 1 can be elminated

$$\cos \theta = \frac{mg}{p\gamma} + \frac{V\rho g}{p\gamma} \qquad [1]$$

where $\dfrac{V\rho g}{p\gamma}$ = buoyance factor

m = mass of slide as measured with electrobalance
g = local gravitational force (979.3 dynes/g)
p = perimeter of the slide (cm)
γ = surface tension of wetting liquid (water = 72.6 ± 0.2 dynes/cm at 20°C)
V = volume of immersed sample at a particular depth
ρ = density of wetting liquid (water = 0.998 g/cc at 20°C)

The angle determined from both these methods is related to interfacial energetics via the Young equation:

$$\gamma_{sa} = \gamma_{sw} + \gamma_{wa} \cos \theta \qquad [2]$$

where γ_{sa} = solid-air interfacial free energy
γ_{sw} = solid-water interfacial free energy
γ_{wa} = water-air interfacial free energy
θ = angle of contact measured through the water phase (Fig. 4)

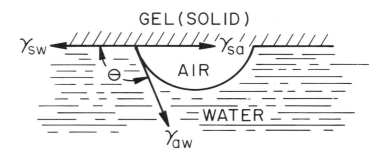

Fig. 4. Captive underwater air bubble contact angle measurement.
γ_{sw}, γ_{aw}, and γ_{sa} represent the solid-water, air-water and
solid-air interfacial free energies.

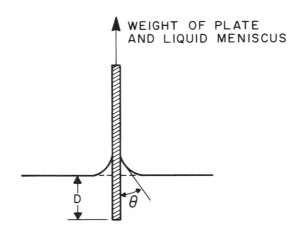

D = DEPTH OF IMMERSION

Fig. 5. Wilhelmy plate technique for contact angle determinations.

The advancing and receding of the water across the polymer surface appears to
involve different mechanisms. This can be observed in the roughness of the advancing angle trace in comparison to the smooth receding angle trace with the Wilhelmy
plate technique for poly(hydroxyethyl methacrylate) surface (Fig. 6). The advancing angle roughness is caused by non-uniform movement of the water over the surface. The water advance involves a rapid zipper-like action from a point of initiation and then stops and repeats the process after the advancing angle reached
a certain critical angle. The receding angle measurement is smooth as a result of
even, uniform movement of the water across the polymer surface.

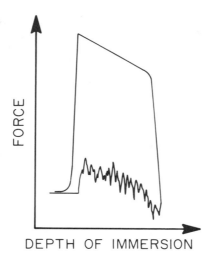

Fig. 6. Wilhelmy plate measurement of poly(hydroxyethyl methacrylate)
[PHEMA] surface.

The large amount of roughness in the advancing angle measurement is not found in
all polymer surfaces. In the HEMA-MMA copolymers, the advancing angle measurement
roughness is minimized at 25 mole % and higher MMA content and the water advancing roughness as shown in Fig. 6 may be considered the exception rather than the
rule. This advancing angle roughness is also observed in some of the alkyl derivatized agarose materials, but only at a specific range of derivatization. The
roughness advancing angle measurement is observed in the HEMA materials even after
prolonged hydration times but is lost from the agarose surface measurements after
hydration.

METHYL METHACRYLATE-HYDROXYETHYL METHACRYLATE COPOLYMERS

The captive bubble and Wilhelmy plate contact angles were measured on surfaces prepared from the same polymer lot. At least three surfaces of the same lot were measured. The Wilhelmy plate measurements were recorded at various times of hydration,
zero time, three hour hydration, 24 hour hydration and after vacuum redrying. These
measurements were recorded as dynamic measurements by immersing the plate and removing it from the water at a speed of 40 mm per minute. At several points in both
the advancing and receding mode of the 24 hour hydrated sample, the plate was
stopped for 15 seconds. The points from the trace were then extrapolated to zero
depth of immersion to obtain the static advancing and receding contact angles. In
most cases, the advancing angle decreases and the receding angle increases, both
slightly; but in some cases, no change is observed between the static and dynamic
contact angles. The captive air bubble measurements were measured only after 24
hour hydration times. These values are shown in Table 4.

From this study, one finds the captive air bubble angles closely follows, but is
slightly higher, than the receding contact angle. It is closer to the static receding angle than the dynamic receding angle and equals this value on the 100%
HEMA surface. The dynamic angle measurements always exhibit more hysteresis than
the static angle measurements except for the very high HEMA content surfaces where
dynamic and static angles are equal. The 24 hour hydrated contact angles are shown

in Fig. 7, along with the equilibrium water content of these copolymers which is essentially linear in regard to the HEMA-MMA mole ratio, ranging from 40% water for the pure HEMA polymer to less than 1% for the MMA polymer. Contact angle hysteresis, the difference between the advancing and receding angle, generally increase with higher HEMA content of the surface, but there are some exceptions. The advancing contact angle exhibits a minimum value at near the 50:50 copolymer ratio; however, the receding angle decreases as the HEMA content of the copolymer increases. As a function of hydration time, the receding angle shows a decrease, whereas the advancing contact angle in general remains constant, or decreases only to a very small extent.

TABLE 4   Comparison of Contact Angles as Determined by Wilhelmy Plate and Captive Underwater Air Bubble Measurements on HEMA-MMA Copolymer Surfaces

Mole % Copolymer	Angles 0 Time	Angles 3 hr. Hyd.	Angles 24 hr. Hyd.	24 hr Hyd. Hyst.	24 hr Hyd. Static Angle	Vacuum Redry	24 hr Hyd. Captive Air Bubbles
100-PMMA	θadv 82.5±1.1	83.7±1.6	83.8±2.2	36°	--	85.2±2.3	59±2
	θrec 55.8±0.2	49.1±0.9	47.8±1.0		--	50.7±1.7	
99-MMA, 1-HEMA	θadv 83.4±2.3	80.4±4.5	82.7±2.6	49.9	74.9±1.7	80.4±4.0	59±1
	θrec 54.8±1.1	51.3±2.0	33.8±1.7		40.9±2.9	48.0±1.3	
95-MMA, 5-HEMA	θadv 80.0±1.7	74.9±2.8	76.6±2.0	32.9	70.6±1.4	76.6±1.0	56±4
	θrec 51.5±1.7	45.2±1.8	43.7±1.0		49.7±1.7	50.5±0.5	
75-MMA 25-HEMA	θadv 73.6±7.4	70.8±0.8	69.9±1.0	38.9	63.0±1.1	72.3±2.3	43±2
	θrec 40.5±0.2	31.9±0.7	31.0±3.1		36.3±2.2	37.7±2.1	
50-MMA, 50-HEMA	θadv 69.1±1.2	64.3±0.7	63.8±0.4	43.2	58.9±0.4	66.6±0.3	32±1
	θrec 28.1±2.0	20.6±1.0	20.6±1.0		26.6±0.7	28.2±1.5	
25 HEMA, 75 HEMA	θadv 65.8±1.4	65.0±1.4	65.9±1.0	56.1	63.8±1.1	71.0±5.7	22±1
	θrec 15.3±0.0	11.6±1.9	9.8±2.7		17.4±2.1	23.8±5.5	
5-MMA, 95-HEMA	θadv 69.7±1.8	72.6±1.5	70.5±3.8	66.0	65.8±4.6	73.9±1.5	18±2
	θrec 13.2±4.1	5.9±3.0	4.0±4.7		12.5±3.0	20.0±1.0	
1-MMA, 99-HEMA	θadv 68.6±0.8	74.5±2.0	65.3±5.2	49.5	65.3±5.3	71.2±0.4	19±2
	θrec 16.8±1.3	10.5±1.9	15.8±3.4		15.8±3.4	14.4±1.5	
100-HEMA	θadv 70.6±0.8	70.8±3.1	69.7±5.0	52.9	69.7±5.0	74.9±1.7	15±1
	θrec 27.0±1.4	13.3±3.4	16.8±1.3		16.8±1.3	13.2±4.1	

hyd. = hydration
hyst. = hysteresis or θadv-θrec

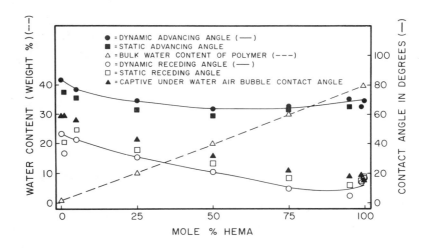

Fig. 7. Contact angle measurements at 24 hour hydration of methyl methacrylate [MMA] and hydroxyethyl methacrylate [MMA] copolymers using both captive air bubble and Wilhelmy plate technique. Also shown is equilibrium water content measurements of the copolymers.

HYDROPHOBIC-HYDROPHILIC-HYDROPHOBIC TRIBLOCK COPOLYMERS

The styrene-hydroxylated butadiene-styrene triblock polymer as with the underivatized polymer exists in domains due to the incompatibility of the polymer blocks (Paul, 1978). All of the polymers were solvent cast from dimethyl formamide except for polystyrene which is cast from toluene. The equilibrium bulk water content and the contact angle results on these polymer systems are shown in Table 3 and Fig. 8. As observed with the HEMA-MMA copolymer study, the captive air bubble underwater contact angle closely corresponds with the dynamic receding contact angle; however, the styrene homopolymer exhibits an anomalous point to this trend with the captive bubble angle nearer to the dynamic advancing angle. This is the only surface we observed where this is found. The receding angle, in contrast to the HEMA-MMA surfaces, does not correspond well with the bulk water content of the polymers. The bulk water content increases almost linearly with butadiene content of the polymer, but the receding contact angle decreases sharply with a small content of hydroborated butadiene and then remains constant as the butadiene content is increased. This may suggest that the underwater surface of all the block polymers is dominated by the hydroborated polybutadiene matrix. The dynamic advancing angle shows a minimum in contact angle at an intermediate butadiene content with both extremes, the polystyrene and hydroborated polybutadiene, having the highest advancing contact angles. The butadiene homopolymer data shown on Fig. 8 is obtained from the hydroborated cis, trans-configuration polymer; however, the contact angle change little from the other hydroborated butadiene configuration.

ALKYL DERIVATIZED AGAROSE SURFACES

n-Butyl or n-dodecyl alkyl groups are covalently bonded to the agarose as shown in Fig. 3. Details of this work along with protein interaction studies will be published elsewhere. The underivatized agarose dissolves in hot water but after a small amount of alkyl group coupling (6.8 μmoles butyl per ml packed gel) it is

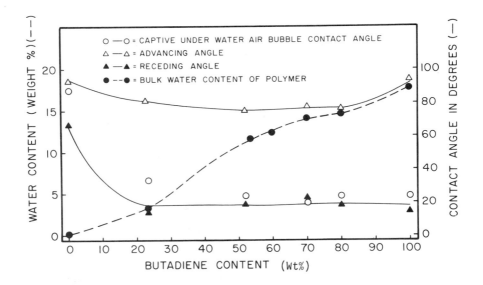

Fig. 8.   Contact angle measurements at 24 hr hydration and equilibrium
water content of styrene-hydroxylated butadiene-styrene triblock copolymers

no longer water soluble. All the materials are readily soluble in dimethyl sulfoxide (DMSO). After solution casting onto clean glass coverslips, the surfaces are
dried in vacuum. The agarose beads contain about 95% water before and after derivatization, but the bulk water content of all the gels decreases substantially when
cast and dried from DMSO. For example, the butyl derivatives and the native agarose swell to about 65% water, but the dodecyl derivatives swelling to a lesser
amount, and the 60.5 μmoles/ml dodecyl agarose only swells to 43% water
(Table 5). For the few agarose materials able to be cast from water, a difference
is shown between the water cast and DMSO cast agarose surfaces and their corresponding contact angles.

The first measurement of the advancing angle on the dried agarose surfaces is
surprisingly hydrophobic but if the slide is immediately redipped to again measure its advancing angle, the angle is changed considerably and immediately becomes more hydrophilic. These hydrated advancing angles become more difficult to
measure because the extrapolated line to zero depth of immersion is many times
non-linear (Fig. 9) due to rapid loss or absorption of water by the gel. Calculation of the second advancing angle as shown in Fig. 9 is in many cases not real-
istic. The advancing and receding contact angles of the dry agarose surface as a
function of degree of derivatization is shown in Fig. 10. The contact angle measurements are shown not to be sensitive to small amounts of alkyl derivatization
in the high water content gels. The 24 hour hydration data is even more difficult
to interpret. The butyl agarose becomes more hydrophobic and generally exhibits
larger contact angles as the degree of derivatization increases, as expected. The
n-dodecyl agarose, however, shows a decrease in contact angle, in other words, the
surface becomes more hydrophilic as the alkyl substitution increases. These samples have been repeated and the same trends have been observed. n-Hexyl and n-
octyl derivatized agarose are presently being prepared to help clarify these results.

TABLE 5 Equilibrium Water Content and Contact Angle Data for Alkyl
Derivatized Agarose Surfaces

Material	Casting Solvent	Degree of Derivatization μmol/ml	Equilibrium Water Content	Dry Surfaces		24 hr. hydrated Surfaces	
				θ Adv	θ Rec	θ Adv	θ Rec
agarose	H₂0	0	—	46.4±3.0	7.7± 1.4	0	0
agarose	DMSO	0	64	68.8±1.8	18.0± 6.5	10.8± 0.5	10.8± 0.5
n-butyl agarose	H₂0	3.9	—	102.2±3.0	9.7± 1.0	82.2± 1.2	8.9± 2.5
n-butyl agarose	DMSO	3.9	61	83.1±1.9	17.6±10.0	8.7±10.3	8.7±10.3
n-butyl agarose	H₂0	6.8	—	117.2±2.7	8.3± 2.2	85.2±5.3	10.1± 9.0
n-butyl agarose	DMSO	6.8	64	109.5±3.3	18.2± 3.1	36.0±3.0	24.6± 5.0
n-butyl agarose	DMSO	13.0	66	98.1±3.1	28.5± 3.1	64.8±6.2	29.9± 5.9
n-butyl agarose	DMSO	19.1	64	92.8±2.5	31.0± 4.0	61.1±3.0	25.6± 2.6
n-butyl agarose	DMSO	31.1	59	92.0±2.4	37.0± 1.9	—	26.2± 7.2
n-dodecyl agarose	DMSO	7.9	60	96.1±1.8	29.5± 1.0	34.6±4.0	34.6± 4.0
n-dodecyl agarose	DMSO	16.1	55	99.2±5.3	26.3± 2.4	27.3±2.0	27.3± 2.0
n-dodecyl agarose	DMSO	38.2	45	100.8±1.9	24.3± 2.4	21.2±4.0	21.2± 4.0
n-dodecyl agarose	DMSO	60.5	43	104.0±1.9	21.3± 3.1	17.8±3.3	17.8± 3.3

SUMMARY AND CONCLUSIONS

Contact angles are extremely useful measurements to determine surface energetics.
The receding angle as measured by Wilhelmy plate procedures most closely approxi-
mates the underwater captive air bubble angle and correlates to the bulk water
content of the systems studied. The receding angle is much more difficult to
interpret and the random and block copolymers exhibit minimum angles at the intermediate (∿ 50-50) copolymer ratios. In the higher water content gels, contact
angle induced surface deformation (Andrade, 1979b) may complicate the analysis.
Contact angle hysteresis is observed in all but the most hydrophilic polymers
studied. Surface roughness is not the cause in the surfaces studied and it has
been shown that roughness is not a serious cause of hysteresis if the rugosities
are less than 0.5 μ (Johnson, 1969). The hysteresis may be due to surface chemical heterogeneity (Johnson, 1964, 1979; Penn, 1980a) In the HEMA-MMA copolymers,
static contact angle hysteresis is usually less than the dynamic contact angle

hysteresis. The angles between the static contact angles in a short time frame (15 sec.) are unchanging provided that solvent evaporation is taken into account and may be due to most stable or metastable intermediate states (Penn, 1980b).

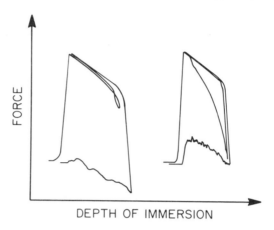

Fig. 9. Wilhelmy plate measurements of agarose surfaces. The surface is initially dry and exhibits a relatively hydrophobic advancing angle. Upon re-immersion, the advancing angle becomes more hydrophilic and in many cases produce non-equilibrium angles which are not readily calculated.

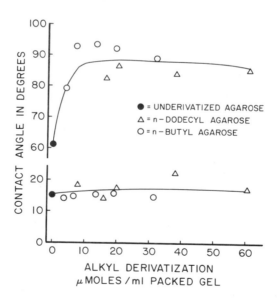

Fig. 10. Advancing (upper line) and receding (lower line) angles n-alkyl derivatized agarose surfaces.

AKNOWLEDGEMENTS

This work was supported by National Institues of Health grants HL26469 and HL-24474. Technical assistance from Ms. Jamil Eghtedari and Ms. Dianne Cress is gratefully acknowledged.

REFERENCES

Andrade, J.D., R.N. King, D.E. Gregonis, and D.L. Coleman (1979) *J. Polym. Sci., Polymer Symposium* 66, 313-336.

Andrade, J.D., S.M. Ma, R.N. King, and D.E. Gregonis (1979) *J. Colloid Interface Sci.*, 72, 488-494.

Bethell, G.S., J.S. Ayers, W.S. Hancock and M.T.W. Hearn (1979) *J. Biol. Chem.*, 254, 2572-2574.

Brown, H.C. and B.C. Subba Rao (1959) *J. Am. Chem. Soc.*, 78, 5694-5695.

Cuatrecasas, P. (1970) *J. Biol. Chem.*, 245, 3059-3065.

Gregonis, D.E., C.M. Chen and J.D. Andrade (1976) in *Hydrogels for Medical and Related Applications*, J.D. Andrade, Ed., ACS Symposium Series 31, American Chemical Society, Washington, D.C., 88-104.

Haller, I. (1978) *J. Am. Chem. Soc.*, 78, 8050-8055.

Hjerten, S., J. Rosengren and S. Pahlman (1974) *J. Chromat.*, 101, 281-288.

Johnson, R.E. and R.H. Dettre (1964) *J. Phys. Chem.*, 63, 1744.

Johnson, R.E. and R.H. Dettre (1969) in *Surface and Colloid Science*, Vol. 2, E. Matijevic, Ed., Wiley-Interscience, New York, 85-153.

King, R.N., J.D. Andrade, S.M. Ma, D.E. Gregonis, and L.R. Brostrom (1981) *J. Colloid Interface Sci.*, accepted for publication.

Levesque, G. and C. Pinazzi (1971) *Bull. Soc. Chim. France*, 3, 1008-1010.

Ochoa, J.-L. (1978) *Biochimie*, 60, 1-15.

Okano, T., J. Aoyagi, and I. Sinohara (1976) *Nippon Kagaku Kaishi*, 1, 161-165.

Patel, K. and W.H. Snyder (1981) *Polymer Preprints*, 22, 217-218.

Paul, D.R. and S. Newman (1978) *Polymer Blends*, Vol. 1, Academic Press, New York.

Penn, L.S. and B. Miller (1980) *J. Colloid Interface Sci.*, 77, 574-576.

Penn, L.S. and B. Miller (1980) *J. Colloid Interface Sci.*, 78, 238-241.

Shaltiel, S. (1974) *Meth. Enzymol.*, 34, 126-140.

Smith, L.M., J.D. Andrade, T. Doyle, and D.E. Gregonis (1981) *J. Appl. Polym. Sci.*, accepted for publication.

Srinivasan, R. and E. Ruckenstein (1980) *Separation and Purification Methods*, 9, 267-370.

Wilhelmy, L. (1863) *Ann. Physik*, 119, 177.

ORGANIC FLOCCULANTS IN DEWATERING FINE COAL AND COAL REFUSE:
STRUCTURE VS. PERFORMANCE

M. E. Lewellyn and S. S. Wang

American Cyanamid Company
Chemical Research Division
1937 West Main Street
Stamford, CT   06904

ABSTRACT

The focus of this paper is the study of the effects of molecular weight and degree of anionicity (as expressed by sodium acrylate content) of an organic flocculant on the vacuum filtration of fine clean coal and coal refuse. For optimum performance, the copolymer compositions containing $\geq$ 70 mole percent of sodium acrylate (the rest being acrylamide) with a low to medium molecular weight (3×10^5 to 5×10^6 depending on acrylate content) are preferred. The results are discussed in terms of current filtration and flocculation theory.

KEYWORDS

Coal; coal refuse; filtration; flocculation; sodium acrylate/acrylamide copolymers; dewatering; synthetic organic flocculant.

INTRODUCTION

New mining techniques introduced into the coal mining industry during the past two or three decades have produced increasing amounts of fines (< 590 μm diameter particles). The amount of fines is expected to increase as coal production increases to meet the energy demand. Recent statistics show that almost 25 percent of the mined coal is rejected and must be disposed of as refuse (Moudgil, 1980). About 170 million tons of refuse is produced each year, of which about 30 million tons is fine material. Disposal of this fine coal refuse is one of the major technical problems of the coal industry. Previously, slurry ponds were used for this disposal, but the increasing amount of such fines and the difficulty in obtaining new permits for ponds has forced producers to look for other means of disposal. Moudgil (1980) has made a comprehensive review of the various methods for disposal. One of the most attractive methods is to use the refuse for land fill which requires efficient dewatering of the fine refuse.

Many years ago, much of the coal fines produced was thrown out with the refuse, but today the economics dictate that as much of the fine clean coal be recovered as possible. It is estimated that in the near future 100 to 150 million tons per year of fines will be handled in coal processing (Sehgal and Clifford, 1980). This fine size fraction of coal retains significantly more moisture than the coarse, and

134

creates problems in meeting moisture specifications of the coal customer. This high moisture content also contributes to a freezing problem in subfreezing weather. Efficient dewatering is needed to alleviate these problems.

A number of methods are used by the coal industry to dewater fine clean coal and coal refuse (Wilson and Miller, 1974). These include thickeners, vacuum filters, pressure filters, belt filters, and centrifuges. In this paper, the focus will be on vacuum filtration and the effects of anionic flocculants in this system.

Dewatering and Filtration Theory

There are many good reviews of the theory of filtration and dewatering (Gray, 1958; Lloyd and Dodds, 1972; Purchas, 1980; Tiller, Crump and Ville, 1980; Vickers, 1981; Wakeman, 1977). Starting with Darcy's law for one-dimensional flow of a fluid through a porous medium, an equation for the residual saturation, S_r, for a filter cake can be derived (Lloyd and Dodds, 1972)

$$S_r = (\frac{1}{86.3} \frac{K}{\gamma \cos \theta} \frac{\Delta P}{L})^{-0.264} \tag{1}$$

where K is the permeability of the bed, γ is the surface tension, θ is the contact angle, ΔP is the pressure drop across the filter cake, and L is the cake thickness. The rate of cake formation can also be derived (Vickers, 1981)

$$\frac{W}{t} = (\frac{2 \, C \, Pa}{\mu \, R_C})^{\frac{1}{2}} \, t^{-\frac{1}{2}} \tag{2}$$

where W is the mass of dry solids deposited per unit area, t is time, C is slurry solids concentration, Pa is the sum of vacuum and average hydrostatic head, μ is viscosity, and R_C is the flow resistance of the cake which is equal to L/K.

It is apparent from equations 1 and 2 that dewatering can be improved by lowering the surface tension and viscosity of the fluid. For coal, this has been amply demonstrated in the literature. Nicol (1976) and Silverblatt and Dahlstrom (1954) have shown that surfactants, which lower surface tension, reduce the cake moisture of a fine coal filter cake. Dahlstrom and Silverblatt (1973) have also shown that steam heat, which primarily lowers the fluid viscosity, lowers the fine coal filter cake moisture. Baker and Deurbrouck (1976) have demonstrated that these effects are additive when both heat and surfactants are used in fine coal filtration. Equation 2 shows that high solids concentration is valuable for good filtration, which means that good thickener performance is important (Matheson and Mackenzie, 1962).

The Kozeny equation (eq. 3) shows that permeability, K, is related to the porosity of the bed, ϵ, and the specific surface of the particles, S (Gray, 1958).

$$K = \frac{\epsilon^3}{S^2 k (1-\epsilon)^2} \tag{3}$$

The Kozeny constant, k, is influenced by particle shape and other factors. Specific surface, S, is the inverse of particle size, thus permeability is directly proportional to the square of the average particle size in the cake. This means that increasing particle size will cause an increase in filtration rate. Mineral slurries, in particular coal refuse slurries, containing a significant amount of ultrafine particles (< 10 μm) are almost impossible to filter. Destabilizing the suspension and aggregating these ultrafine particles can improve filtration significantly. This has been well demonstrated with the use of flocculants in filtration of coal and coal refuse (Geer, Jacobsen, and Yancey, 1959; Gieseke, 1962; Gray, 1958;

Matheson and Mackenzie, 1962; Mishra, 1973; Osborne and Robinson, 1973; Pearse and Barnett, 1980; Rushton, 1976). Originally, the flocculants used in coal processing were naturally occuring polymers such as starch and gelatine. Synthetic polyelectrolytes, introduced in the 1950's, were found to be much more effective flocculating agents than the natural polymers (Geer, Jacobsen, and Yancey, 1959; Gray, 1958), and are now used extensively in the coal industry.

Flocculants and Flocculation

Although there are several hundred commercially different flocculants available, there are a limited number of chemically different types available. Synthetic polyelectrolyte flocculants are available in three basic charge types with a wide range of molecular weights and varied charge densities (Halverson and Panzer, 1980; Werneke, 1979). The non-ionic flocculants are almost exclusively polyacrylamide-based materials. The anionic flocculants are usually copolymers of acrylamide and acrylic acid (or sodium acrylate) or partially hydrolyzed polyacrylamide. Cationic polymers generally contain a quaternary ammonium group. These can be obtained from copolymers of acrylamide and a suitable quaternized monomer, condensation of two monomers such as epichlorohydrin and dimethylamine, or by homopolymerization of a quaternary ammonium monomer such as diallyldimethylammonium chloride.

There are a number of excellent reviews on the subject of flocculation mechanisms (Akers, 1976; Halverson, 1978; Halverson and Panzer, 1980; Somasundaran, 1980). The two basic mechanisms for flocculation with synthetic polymers are the charge patch mechanism and the bridging mechanism. The charge patch mechanism generally involves the adsorption of a cationic polymer onto a particle with negatively charged sites via electrostatic bonds. The cationic polymer covers these negative sites and effectively neutralizes the particle surface. The cationic patch thus formed provides an area for electrostatic bonding with the negative surface of another particle.

In the bridging mechanism, the polymer chain is adsorbed on the particle surface at only a few points of attachment, leaving either ends or loops extending into the solution for contacting other particles. The bonding of the polymer to the particle surface is by a variety of mechanisms. These include electrostatic bonding, hydrophobic bonding, Van der Waals bonding, and a group of rather specific physico-chemical interactions ranging from hydrogen bonding to covalent bonding.

For filtration there is an optimum dosage beyond which the filtration performance deteriorates (Linke and Booth, 1960; Mishra, 1973; Pearse and Barnett, 1980; Rushton, 1976). It has been suggested by various authors that the maximum filtration rate occurs when the fraction of particle surface covered by the polymer molecules is close to 0.5, but this has not been experimentally proven (Somasundaran, 1980). Overdosing can lead to restabilization of the slurry and thus blinding with fine particles. Overdosing can also lead to greater flow resistance of the filter media (Rushton, 1976).

The type of floc formed is also important for filtration. Small tight flocs are preferred over large loose flocs (Pearse and Barnett, 1980; Werneke, 1979). Large loose flocs contain entrapped water which leads to high moisture contents and they tend to collapse on the filter medium blinding the medium.

Although there has been much work that shows that flocculants improve the filtration of fine coal and coal refuse, very little has been reported on the correlation of structure with performance. It is generally accepted that anionic flocculants are best (Osborne and Robinson, 1973; Pearse and Barnett, 1980; Werneke, 1979), but little is known about the best structure needed for filtration. This paper will focus on the relationship of carboxylate content and molecular weight of anionic flocculants to their filtration performance with fine coal and coal refuse.

RESULTS AND DISCUSSION

There are a number of factors that can influence filtration (Matheson and Mackenzie, 1962; Nelson and Dahlstrom, 1957). These include, in addition to those already mentioned, pH, vacuum level, form and dry times, and filter media. The experimental method used for most of the tests was the standard filter leaf test (Dahlstrom and Silverblatt, 1977). All factors other than the amount and type of polymer used were kept as constant as possible, with a given substrate, so that the effect of polymer structure on filtration could be determined. The slurries were filtered at their natural pH (7.5-8.0), which is the pH normally used in coal preparation. The substrates were obtained as slurries from coal preparation plants in the Appalachian region. The type of substrate and physical properties are listed in Table 1. The polymers used for these tests were copolymers of acrylamide and sodium acrylate. These polymers are listed in Table 2 with the mole percent of sodium acrylate content and average molecular weight. The number used to identify the polymer is a combination of acrylate content and molecular weight for ease of identification in the discussion.

TABLE 1 Substrate Properties

Type of Substrate		A Thickener Underflow	B Thickener Underflow	C Flotation Concentrate	D Flotation Concentrate
	> 590 μm	1.3	1.8	-	22.6
Percent Particle Size Distribution	149-590 μm	12.9	17.6	4.8	39.3
	74-149 μm	6.7	9.1	10.6	15.2
	44-74 μm	4.2	5.7	9.1	6.6
	< 44 μm	74.8	65.8	75.5	16.3
Percent Ash Content		63.3	64.7	13.4	12.2
Percent Solids		26	30	20	35

TABLE 2 Polymer Characteristics

Polymer	Mole % Acrylate	Estimated Average Molecular Weight
30-40	30	4×10^6
30-120	30	1.2×10^7
50-8	50	8×10^5
50-25	50	2.5×10^6
70-3	70	3×10^5
70-5	70	5×10^5
70-8	70	8×10^5

TABLE 2   Polymer Characteristics (continued)

Polymer	Mole % Acrylate	Estimated Average Molecular Weight
70-25	70	2.5×10^6
70-45	70	4.5×10^6
85-8	85	8×10^5
85-25	85	2.5×10^6
95-50	95	5×10^6
95-90	95	9×10^6
95-120	95	1.2×10^7
95-150	95	1.5×10^7
95-180	95	1.8×10^7
100-50	100	5×10^6
100-70	100	7×10^6
100-120	100	1.2×10^7
100-150	100	1.5×10^7
C[1]	-	7×10^4

[1]Copolymer of dimethylamine and epichlorohydrin

Coal Refuse

The coal refuse substrates used for this study are substrates A and B in Table 1.
The noncoal content of coal refuse is mainly clays, the majority being illitic and
kaolinitic clays for Appalachian coal (Bradley, Aplan, and Hogg, 1979). These substrates are very difficult to filter due to the large amount of ultrafine particles.
A Coulter Counter analysis of the particle size portion less than 44 μm for substrate A gave a geometric mean particle size of 7.7 μm.

A convenient screening method for polymers to determine optimum dosage with the
minimum of substrate was the Buchner funnel test. With substrate A, a number of
different anionic flocculants were examined along with a cationic for their filtration efficiency (Fig. 1). Anionic flocculants gave the best performance of all
those tested, although this performance varied greatly with polymer structure.

The results show that the better flocculants were those of high sodium acrylate
content (50-85%) and low molecular weight (8×10^5) with a minimum filter time of
6-10 minutes. Flocculants 30-40 and 30-120 are typically used in coal refuse
thickeners (Werneke, 1979), but they are inferior reagents for vacuum filtration,
with a minimum filter time of 20-22 minutes. The effect of molecular weight is
clearly shown with these two flocculants, the lower molecular weight product
(4×10^6) requiring only 100-150 gm/ton for the fastest filtration as compared to
200 gm/ton for 30-120 with a molecular weight of 1.2×10^7. This also shows the
detrimental affect of overdosing in that both 30-40 and 30-120 demonstrate a definite minimum in their dosage curves. The cationic polymer, C, was not very effective in increasing the filtration rate in the dosage range tested when used by it-
self. Cationic polymers are reported to be beneficial when used in combination
with an anionic polymer (Werneke, 1979). It should also be noted that the high
acrylate containing polymers (70-8 and 85-8) show a broader range of optimum dosage
(150-250 gm/ton) than do the ones with only 30% acrylate (30-40 and 30-120).

Since the better performance in the Buchner funnel tests was obtained with high
acrylate containing polymers, a series of polymers containing 50-85 mole percent
sodium acrylate with a molecular weight range of 3×10^5 to 4.5×10^6 were examined

with substrate A. The results of these tests are summarized in Table 3. The results shown are the averages of at least two tests for each data point using the
standard filter leaf method. The cake weights given are for an area of 0.0093 m²,
the area of a standard test leaf. The dosage of 150 gm/ton was used since this was
at the lower end of the optimum range as shown in Fig. 1. The use of this dosage
as optimum for comparison is confirmed by the dosage study for 70-8 using a filter
leaf (Table 4). The other polymers showed similar results.

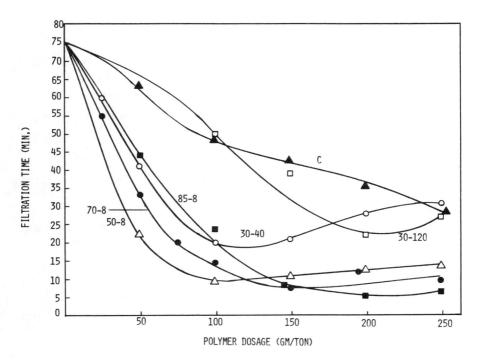

Fig. 1. Filtration time vs. dosage for coal refuse
substrate A measured with Buchner funnel.

TABLE 3   Filter Leaf Test Results with Coal Refuse
Substrate A

Polymer	Dosage (gm/ton)	Cake Weight (gm)	Percent Moisture
None	-	7.3	40.9
50-8	150	23.6	34.9
50-25	150	20.0	35.3
70-3	150	29.3	33.2
70-5	150	29.5	34.4
70-8	150	29.5	34.5
70-25	150	24.5	33.0
70-45	150	18.6	34.5
85-8	150	27.7	33.1
85-25	150	24.9	34.8

TABLE 3   Filter Leaf Test Results with Coal Refuse
          Substrate A (continued)

Polymer	Dosage (gm/ton)	Cake Weight (gm)	Percent Moisture
95-50	150	47.6	35.7
95-120	150	46.3	33.9
95-180	150	48.5	34.7

TABLE 4   Dosage Study for 70-8 with Coal Refuse
          Substrate A by Filter Leaf

Dosage (gm/ton)	Cake Weight (gm)	Percent Moisture
50	16.0	33.9
100	23.8	33.2
150	29.5	34.5
250	25.4	35.1

At a dosage of 150 gm/ton, the moisture content ranged from 33.0% to 35.3% and the
filter cake weight ranged from 18.6 gm to 29.5 gm as compared to a moisture content
of 40.9% and a cake weight of 7.3 gm for the untreated coal refuse. To best see
the effects of the polymer structure on filtration, these results are analyzed
according to sodium acrylate content (Fig. 2 and Fig. 3) and molecular weight
(Fig. 4 and Fig. 5).

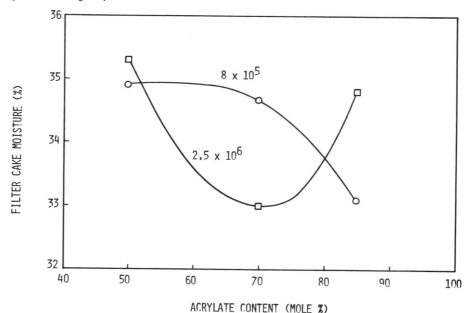

Fig. 2.   Effect of acrylate content on coal refuse substrate
          A filter cake moisture with molecular weights of
          8×10^5 and 2.5×10^6 at a dosage of 150 gm/ton.

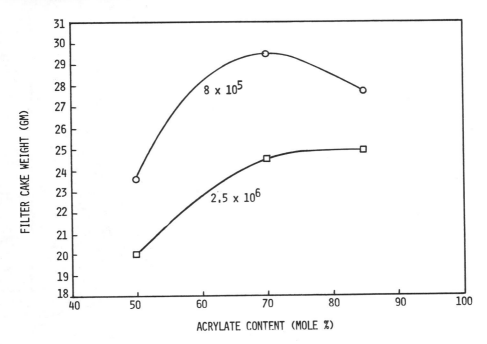

Fig. 3.   Effect of acrylate content on coal refuse substrate
A filter cake weight with molecular weights of
8×10^5 and 2.5×10^6 at a dosage of 150 gm/ton.

In comparing filter cake moisture with sodium acrylate content of the polymer, no
clear cut trend was observed (Fig. 2). For a molecular weight of 8×10^5 there is
a definite trend to lower moisture with increasing sodium acrylate content. However, at a molecular weight of 2.5×10^6, the data show a minimum at 70% acrylate
content with the moisture content at both 50% and 85% acrylate content being
about 2% higher. One observation can be made and that is that the polymers with
50% acrylate content produced moisture contents as high as and generally higher
than those with 70% and 85%.

Although there was not a definite trend with moisture content, the comparison of
filter cake weight with acrylate content shows a clear pattern (Fig. 3). At both
molecular weights, the 70% and 85% acrylate containing polymers gave about equal
cake weights which were significantly (17% to 25%) greater than that of 50%
acrylate. It should also be noted that the curve for the 8×10^5 molecular weight
is higher (~ 4 gm) than that for 2.5×10^6 molecular weight.

As with acrylate content, it is difficult to see any general trend in comparing the
moisture content with polymer molecular weight (Fig. 4). At 50% acrylate content,
both molecular weights gave about equal moisture contents of about 35%; and at 85%
acrylate content, the moisture content increases from 33% at a molecular weight of
8×10^5 to about 35% at a molecular weight of 2.5×10^6. However, the 70% acrylate
containing polymers show no trend, varying from 33% to 34.5% throughout the molecular weight range of 3×10^5 to 4.5×10^6.

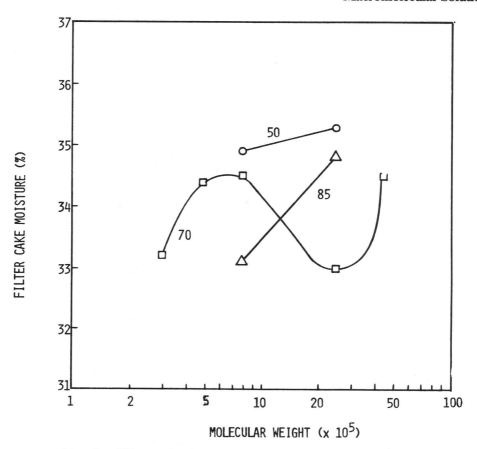

Fig. 4. Effect of molecular weight on coal refuse substrate
          A filter cake moisture with acrylate contents of
          50%, 70%, and 85% at a dosage of 150 gm/ton.

Again, in contrast to the moisture content data, there is a pronounced trend in
comparing the filter cake weight with molecular weight (Fig. 5). With all acrylate
contents, the cake weight decreases with increasing molecular weight. This trend
is most pronounced with 70% acrylate content which has the widest molecular weight
range (3×10^5 to 4.5×10^6). With this series the molecular weights of 3×10^5
to 8×10^5 gave about equal results (about 29 gm), but the performance drops off
rapidly to 24.5 gm at 2.5×10^6 and 18.6 gm at 4.5×10^6.

In summary, for this series of polymers, the best performance of high cake weight
and low moisture content is obtained with a sodium acrylate content of 70% to 85%
and a molecular weight of 3×10^5 to 8×10^5.

Since the best results were obtained with the polymers of higher sodium acrylate
content in this series, another series of polymers were examined with 95% sodium
acrylate content (95-50, 95-120, and 95-180 in Table 3). The average molecular
weights for these polymers ranged from 5×10^6 to 1.8×10^7. These gave markedly
higher cake weights than did the first series (46.3 to 48.5 gm, 57% to 65% greater

than the best of the first series). The moisture content was about the same as
with the 50% to 85% acrylate polymers, the lowest (33.9%) being obtained with the
1.2×10^7 molecular weight polymer. It is obvious that these polymers are by far
the superior flocculants for coal refuse substrate A.

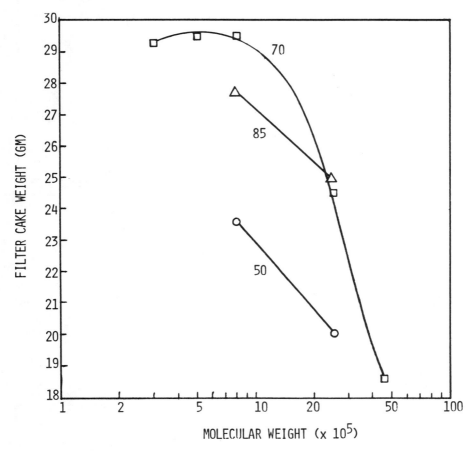

Fig. 5. Effect of molecular weight on coal refuse substrate
A filter cake weight with acrylate contents of 50%,
70%, and 85% at a dosage of 150 gm/ton.

The filtration performance of these polymers was examined with another coal refuse
substrate (B in Table 1) in order to confirm the trends observed with substrate A,
and the results are summarized in Table 5. This is a somewhat coarser substrate
which accounts for its better filtration performance without treatment. Although
little improvement in moisture content was obtained by treatment with polymers,
significant increases in filter cake weight were observed. The best performance
was again obtained with polymers with high sodium acrylate content. Molecular
weight exhibits a definite effect on cake weight. With the 30% acrylate polymer,
the lower molecular weight product (4×10^6) was better than the higher molecular
weight product (1.2×10^7). A similar result was obtained with substrate A (Fig. 1).

Unlike substrate A, however, the 95% acrylate polymers exhibit a pronounced molecular effect with the lower molecular weight product (5×10^6) providing the best filtration performance over that of the higher molecular weight products (1.2×10^7 and 1.8×10^7).

TABLE 5 Filter Leaf Test Results with Coal Refuse Substrate B

Polymer	Dosage (gm/ton)	Cake Weight (gm)	Percent Moisture
None	-	17.0	34.8
30-40	150	33.3	35.4
30-120	150	26.8	35.7
70-8	150	36.1	33.5
95-50	150	45.6	33.7
95-120	150	37.2	33.6
95-180	150	32.8	33.2

In summary, at least for these two coal refuse substrates, the best flocculants for vacuum filtration are those containing a high sodium acrylate content and with low to moderate molecular weight. Of those tested, the best is 95-50 with a sodium acrylate content of 95% and a molecular weight of 5×10^6.

Clean Coal

The results obtained with fine coal refuse were examined with clean coal substrates to determine if the trends are similar. The substrates used were C and D in Table 1. There is a large difference between the two substrates as to the particle size distribution. Substrate C has a significantly finer particle size (75.5% < 44 µm) than substrate D (16.3% < 44 µm). The ash contents are similar (12-13%), the majority of the ash being clays similar to the coal refuse.

The filter leaf test results for substrate C are summarized in Table 6. The dosages of polymer needed are lower than that for coal refuse and the resulting cake weights are much greater. All of the flocculants produced a significant reduction in moisture content (40.2% down to 32-35%). This is shown in Fig. 6 for a few of the polymers. However, there are differences in filter cake weight, particularly at the higher dosage of 100 gm/ton (Fig. 7). The two polymers with 30% acrylate content (30-40 and 30-120) show a maximum cake weight at 50 gm/ton and then either levels off or decreases at 100 gm/ton. The polymers containing 95% and 100% acrylate content (95-50 and 100-50), while producing only slightly higher cake weights at 50 gm/ton than the 30% acrylate polymers, improve to a significantly higher cake weight at 100 gm/ton. It is interesting to note that the low molecular weight (3×10^5) polymer 70-3 (with 70% acrylate content) performed well in comparison to the higher molecular weight polymers.

As seen with the coal refuse, there are some molecular weight effects on polymer performance. At a dosage of 50 gm/ton, the lower molecular weight 30% acrylate containing polymer (30-40) produces a notably heavier filter cake (93.1 gm vs. 73.1 gm) than the higher molecular weight polymer (30-120). The same type of molecular weight dependence is shown for the 95% and 100% acrylate content polymers (Fig. 8). The data clearly show that the better performance, as far as cake weight is concerned, is obtained with the lower molecular weight products (5×10^6). However, too low a molecular weight is detrimental as shown in another series of tests for the 70% acrylate content polymers (Table 7). In this case, the polymers with molecular weights of 8×10^5 and 2.5×10^6 produced significantly higher cake weights than the lower molecular weight polymer (3×10^5).

TABLE 6 Filter Leaf Test Results with Clean Coal Substrate C

Polymer	Dosage (gm/ton)	Cake Weight (gm)	Percent Moisture
None	-	27.5	40.2
30-40	50	93.1	33.7
30-40	100	75.2	34.1
30-120	50	73.1	32.9
30-120	100	78.4	32.1
70-3	25	64.3	35.5
70-3	50	83.0	34.5
70-3	100	89.1	34.4
95-50	50	99.6	33.9
95-50	100	120.9	33.2
95-90	50	96.7	33.4
95-90	100	113.9	34.7
95-120	50	89.5	33.8
95-120	100	108.3	33.4
95-150	50	80.9	33.9
95-150	100	98.5	34.7
95-180	50	70.6	35.6
95-180	100	92.0	33.6
100-50	50	98.8	33.3
100-50	100	112.9	34.4
100-70	50	88.7	32.9
100-70	100	113.3	32.8
100-120	50	73.8	36.9
100-120	100	99.2	35.5
100-150	50	70.4	35.6
100-150	100	95.9	34.8

TABLE 7 Filter Leaf Test Results with Clean Coal Substrate C for 70% Acrylate Content Polymers at a Dosage of 100 gm/ton

Molecular Weight (x 10^5)	Cake Weight (gm)	Percent Moisture
3	89.2	33.3
8	123.2	32.4
25	110.3	33.1

Although substrate D contains a considerably coarser particle dize distribution than substrate C, the small amount of fines present is enough to noticeably affect the filtration performance in an adverse way (Table 8). Only a very small amount of polymer (25 gm/ton) is needed to greatly improve this filtration (Table 8).

As with substrate C, there is a marked effect of molecular weight on polymer performance for filter cake weight. However, for the 30% acrylate containing polymers the effect is the opposite. The better performance was obtained with the higher molecular weight polymer 30-120 (103.1 gm vs. 58.5 gm cake weight). The molecular weight effect of the 95% acrylate containing polymers is the same as for substrate D, but with more significant differences. In this case the cake weight decreases with increasing molecular weight from 109.2 gm (5 x 10^6 molecular weight) to only 19.4 gm (1.8 x 10^7 molecular weight).

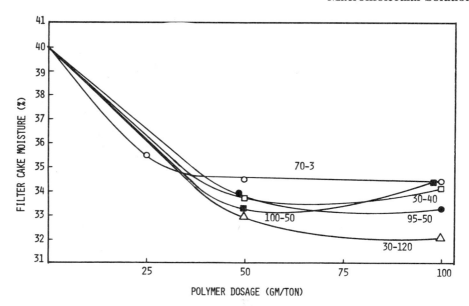

Fig. 6. Filter cake moisture vs. dosage for clean coal
         substrate C.

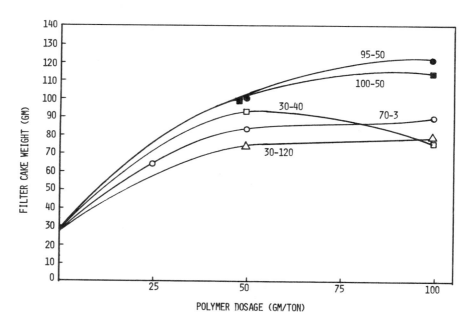

Fig. 7. Filter cake weight vs. dosage for clean coal
         substrate C.

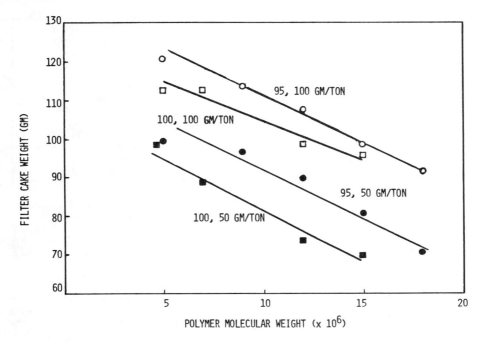

Fig. 8. Filter cake weight vs. molecular weight for clean
coal substrate C with 95% and 100% acrylate content
at polymer dosages of 50 and 100 gm/ton.

TABLE 8   Filter Leaf Test Results with Clean Coal Substrate D

Polymer	Dosage (gm/ton)	Cake Weight (gm)	Percent Moisture
None	-	13.4	31.1
30-40	25	58.5	26.8
30-120	25	103.1	25.6
70-8	25	71.4	30.9
95-50	25	109.2	25.4
95-120	25	60.0	32.2
95-180	25	19.4	27.8

The data clearly show that, for these two clean coal substrates, the high acrylate
containing polymers perform very well in vacuum filtration as they do also for fine
coal refuse. The best polymer is again 95-50, which generally produced the highest
cake weights. As with the coal refuse substrates, there is a definite dependence
of filtration performance on molecular weight, the best performance being with the
lower molecular weight products (5×10^6 for 95% acrylate content).

CONCLUSIONS

A number of studies have shown that for settling of clays, the optimum acrylate
content is 20-30% (Halverson, 1979; Pearse and Barnett, 1980; Werneke, 1979).

However, what is good for settling is not necessarily good for filtration since large loose flocs are desirable for fast settling, but small tight flocs are better for good vacuum filtration. This study has shown that high acrylate containing polymers (> 70%) provide the best filtration performance.

The mechanism for this improved filtration performance is not clear. Since the clay particles normally exhibit a negative charge at the pH at which these tests were conducted (pH 7.5-8.0) (Halverson, 1978; Pearse and Barnett, 1980), one would ex-pext the highly anionic polymers to be repulsed. This is not the case, however, as seen by the fact that excellent filtration is obtained with the high acrylate content. Since the water used in coal preparation plants is recycled, the electro-lyte content is high and contains a significant amount of calcium and magnesium ions. For example, the water in substrate A contains 55 ppm of calcium and 30 ppm of magnesium measured as the carbonates. Although the high acrylate containing polymers are almost completely ionized at the pH used for the tests, the effective negative charge would be less due to binding of cations to the polymer (Miller, 1964). Polyvalent cations, such as calcium and magnesium, are bound more tightly than sodium and thus their presence in solution would effectively neutralize much of the negative charge and reduce the expected repulsion. This tight binding of the polymer with polyvalent cations would provide a means for good bonding of the polymer to the particles through a polyvalent cation bridge (Matheson and Mackenzie, 1962). Since the polymer is almost completely ionized at this pH, it would be at maximum extension (although not necessarily completely extended). Data on poly-methacrylic acid, in the absence of added electrolyte, show that at $\geq$ 50% ionization, the maximum polymer size was obtained which was about seven times the size for the unneutralized polymer (Oth and Dotz, 1952). The expansion is not as pronounced in the presence of electrolytes. This expansion of the polymer would explain the mole-cular weight effect on filtration performance. Since the polymer is extended a lower molecular weight is required for the necessary length to bridge particles. The higher molecular weight polymers will produce looser flocs due to the longer chain length and charge repulsion of the particles. These loose flocs are detri-mental to filtration in that they entrap water and collapse on the filter medium, blinding it. With the low molecular weight polymers of high acrylate content, bridging does occur, but the floc size would be kept small due to charge repulsion of the particles and shorter length of the polymer.

In most coal preparation plants, the same flocculant used in their thickeners (gen-erally an anionic containing 30% acrylate) is also used in the filtration of clean coal and coal refuse. It is evident from this work that significant improvements in throughput can be obtained by using a polymer containing > 70% acrylate, in par-ticular one containing 95% acrylate. In determining the best reagents for filtra-tion in a coal preparation plant, other polymers than those used normally in thick-ening should be examined to obtain the best performance.

EXPERIMENTAL

The polymers used in these tests were obtained from others at this laboratory using standard methods for the copolymerization of acrylamide and sodium acrylate (see acknowledgements). The molecular weights given are only approximate and are based on viscosity studies.

Filter Leaf Test

The standard filter leaf test as outlined by Dahlstrom and Silverblatt (1977) was used. For substrates A and D, in order to obtain the maximum data from the avai-lable amount of slurry, a smaller leaf of 0.0033 m^2 was used. This was made by removing the glass side from a 150 ml size coarse Pyrex fitted glass filter. A 7 cm Whatman No. 541 filter paper was used as the filter medium. The slurry was weighed into a 600 ml beaker and was treated with the polymer (as a 0.1% aqueous solution) while stirring with a 2 in. three blade propeller type stirrer at 500 rpm. The slurry was conditioned for 1 minute and the stirrer removed. Using a vacuum of 68 kPa, the leaf was immersed into the slurry, using a slow stirring motion. For substrate A, a form time of 1.5 minutes was used and for substrate D a form time of 0.5 minutes was used. At the end of the form time, the leaf was rotated out of the slurring to an upright position and air was pulled through the cake. For substrate A the dry time was 1.5 minutes and for substrate D it was 1 minute. At the end of the dry time, the cake was removed and the moisture determined by oven drying overnight at 65°C. For convenience of comparison, the cake weights were adjusted to that for an area of 0.0093 m^2. With substrates B and C, the standard 0.0093 m^2 filter leaf was used with polypropylene filter cloth. The slurry was weighed into a 3 liter beaker and the same procedure was used. The form times and dry times for both substrates were 1.5 minutes. For each set of conditions, at least two tests were conducted and the results averaged. The spread between data points was small.

Buchner Funnel Test

Into a 250 ml beaker was weighed 154 gm of substrate A. The slurry was treated with polymer, while stirring, in the same manner as for the filter leaf test. The slurry was poured into a 7 cm Buchner funnel, using a 7 cm Whatman No. 541 filter paper, with a vacuum of 68 kPa. The time needed for the disappearance of the water from the top of the cake was recorded.

ACKNOWLEDGEMENTS

The authors wish to express their gratitude to Dr. R. Neff and Dr. R. Nahas for providing the polymers used in these studies. Thanks are also due to Mr. C. Dugan and Mr. J. Dobson for their technical assistance. Recognition is due to Dr. F. Halverson for his advice.

REFERENCES

Akers, R. (1976). Flocculation. Rep. Prog. Appl. Chem., 60, 605-621.
Baker, A. F., and A. W. Deurbrouck (1976). Hot surfactant solution as dewatering aid during filtration. In A. C. Partridge (Ed.), Proceedings of the Seventh International Coal Preparation Congress, B.2., 21 pp.
Bradley, P. B., F. F. Aplan, and R. Hogg (1979). Characterization of solid constituents in blackwater effluents from coal preparation plants. Interagency Energy/ Environment R & D Program Report Fe-9002-1, EPA-600/7-79-006, 203 pp.
Dahlstrom, D. A., and C. Silverblatt (1973). Production of low moisture content fine coal without thermal drying. Min. Cong. J., 59, No. 12, 32-40.
Dahlstrom, D. A., and C. E. Silverblatt (1977). Continuous vacuum and pressure filtration. In D. B. Purchas (Ed.), Solid/Liquid Separation Equipment Scale-Up, Uplands Press Ltd., Croydon, England. pp. 445-491.
Geer, M. R., P. S. Jacobsen, and H. F. Yancey (1959). Flocculation to improve coal slurry filtration. Mining Eng., 11, 715-719.
Gieseke, E. W. (1962). Flocculants and filtration of coal flotation concentrates and tailings. Trans. AIME, 223, 352-358.
Gray, V. R. (1958). The dewatering of fine coal. J. Inst. Fuel, 31, 96-108.
Halverson, F. (1978). Selected physical-chemical aspects of flocculation. Proc. Tenth Annual Meeting, Canadian Mineral Processors, Ottawa, Can., Jan. 1978, pp. 404-450.

Halverson, F., and H. P. Panzer (1980). Flocculating Agents. In M. Grayson and
    D. Eckroth (Ed.), Kirk-Othmer: Encyclopedia of Chemical Technology, Vol. 10,
    3rd ed. John Wiley and Sons, Inc., New York. pp. 489-523.
Linke, W. F., and R. B. Booth (1960). Physical chemical aspects of flocculation
    by polymers. Trans. AIME, 217, 364-371.
Lloyd, P. J., and J. A. Dodds (1972). Liquid retention in filter cakes. Filtr.
    Sep., 9, 91-96.
Matheson, G. H., and J. M. W. Mackenzie (1962). Filtration of flocculated coal
    concentrates containing expanding lattice clays. Trans. AIME, 223, 167-172.
Miller, M. L. (1964). Acrylic acid polymers. In H. F. Mark, N. G. Gaylord, and
    N. M. Bikales (Ed.), Encyclopedia of Polymer Science and Technology, Vol. 1.
    John Wiley and Sons, Inc., New York. pp. 197-226.
Mishra, S. K. (1973). Effect of flocculation on moisture reduction of fine coal.
    Coal Min. Proc., October, 56-59.
Moudgil, B. M. (1980). Handling and disposal of coal preparation plant refuse.
    In P. Somasundaran (Ed.), Proceedings of the International Symposium on Fine
    Particles Processing, Las Vegas, Nevada, February 24-28, 1980, Vol. 2. American
    Institute of Mining, Metallurgical, and Petroleum Engineers, Inc., New York,
    pp. 1754-1779.
Nelson, P. A., and D. A. Dahlstrom (1957). Moisture-content correlation of rotary
    vacuum filter cakes. Chem. Eng. Prog., 53, 320-327.
Nicol, S. K. (1976). The effect of surfactants on the dewatering of fine coal.
    Proc., Australas. Inst. Min. Metall., 260, 37-44.
Osborne, D. G., and H. Y. Robinson (1973). Improving the efficiency of rotary
    vacuum filtration of coal slurries. Filtr. Sep., 10, 153-161.
Oth, A., and P. Doty (1952). Macro-ions. II. Polymethacrylic acid. J. Phys.
    Chem., 56, 43-50.
Pearse, M. J., and J. Barnett (1980). Chemical treatments for thickening and fil-
    tration. Filtr. Sep., 17, 460-470.
Purchas, D. B. (1980). A practical view of filtration theory. Filtr. Sep., 17,
    147-151.
Rushton, A. (1976). Liquid-solid separation - recent research evaluated. Filtr.
    Sep., 13, 573-578.
Sehgal, R. S., and K. L. Clifford (1980). Where are we going with fine coal?
    Presented at SME-AIME Fall Meeting, Minneapolis, MN, October, SME Preprint No.
    80-351, 7 pp.
Silverblatt, C. E. and D. A. Dahlstrom (1954). Moisture content of a fine-coal
    filter cake. Ind. Eng. Chem., 46, 1201-1207.
Somasundaran, P. (1980). Principles of flocculation, dispersion, and selective
    flocculation. In P. Somasundaran (Ed.), Proceedings of the International Sympo-
    sium on Fine Particles Processing, Las Vegas, Nevada, February 24-28, 1980,
    Vol. 2. American Institute of Mining, Metallurgical, and Petroleum Engineers,
    Inc., New York. pp. 947-976.
Tiller, F. M., J. R. Crump, and F. Ville (1980). A revised approach to the theory
    of cake filtration. In P. Somasundaran (Ed.), Proceedings of the International
    Symposium on Fine Particles Processing, Las Vegas, Nevada, February 24-28, 1980,
    Vol. 2. American Institute of Mining, Metallurgical and Petroleum Engineers, Inc.,
    New York. pp. 1549-1582.
Vickers, F. (1981). Filtration theory applied to vacuum filtration in coal prepa-
    ration. Filtr. Sep., 18, 46-52.
Wakeman, R. J. (1977). Cake dewatering. In L. Svarovsky (Ed.), Solid-Liquid
    Separation, Butterworths, London. pp. 297-306.
Werneke, M. F. (1979). Application of synthetic polymers in coal preparation.
    Presented at AIME Annual Meeting, New Orleans, LA, February, SME Prepring No.
    79-106, 11 pp.
Wilson, E. B., and F. G. Miller (1974). Coal dewatering - some technical and
    economic considerations. Min. Cong. J., 60, No. 9, 116-121.

EFFECT OF POLYMER-SURFACTANT INTERACTIONS ON POLYMER SOLUTION PROPERTIES

P. Somasundaran and B. M. Moudgil*

Henry Krumb School of Mines, Columbia University
New York, NY   10027

ABSTRACT

Modifications in the polymer solution properties occur, in addition to changes in the solvent power of the medium also, due to associative interactions between polymer and surfactant species in the bulk. In this investigation effect of polymer (nonionic, anionic and cationic polyacrylamide) and sodium dodecylsulfonate or dodecylaminehydrochloride on polymer solution properties such as relative viscosity, conductivity, surface tension and precipitation behavior has been studied. Characteristics of both the surfactant and the polymer are found to exert significant influence on bulk and interfacial properties. The data obtained is analyzed to elucidate the mechanisms underlying the interactions between various species and their effect on polymer solution properties.

KEYWORDS

Polymer-surfactant interaction; nonionic/ionic polyacrylamide solution properties; precipitation-redissolution in polymer and in surfactant systems; polymer and surfactant surface tension; conductivity; relative viscosity; solubility.

INTRODUCTION

Interactions between polymer and surfactant can be of importance in a wide variety of systems such as in mineral processing and in enhanced oil recovery. In mineral processing, interactions between surface active collector species and long chain polymeric molecules are encountered (i) in Floc Flotation which involves separation of selectively flocculated material from the mineral in suspension using the froth flotation technique and (ii) when polymeric reagents are employed as "depressants" in enhancing the selectivity of flotation separation of minerals. Selectivity in mineral processing operations can be affected through modifications caused by these interactions in the polymer solution properties as well as, in interfacial properties such as adsorption of different species at the solid/liquid and liquid/gas interfaces (Somasundaran, 1969). In this work the role of the

---

* Present Address: Department of Materials Science and Engineering, University of Florida, Gainesville, Fl.,   32611

changes in polymer solution properties is studied in detail for selected polymer surfactant combinations. Specifically the effect of interactions between charged and uncharged polyacrylamides and sodium dodecylsulfonate or dodecylamine hydrochloride, on the polymer solution properties such as surface tension, relative viscosity, conductivity and precipitation behavior is considered in this paper.

EXPERIMENTAL

Polymers: C-14 labeled nonionic (PAM) and ionic sulfonated acrylamide (PAMS) or aminiated acrylamide (PAMD) based polymers were synthesized using radiation induced heterogeneous polymerization technique (Wada, Sekiya and Machi, 1975, 1976). This polymerization technique was selected over other methods because in the present study it is required that polymers not be contaminated by surfactants and other modifiers. A description of the synthesis and characterization technique has been given elsewhere (Moudgil, 1981).

Anionic copolymers (PAMS) were synthesized using 3 mol % of 2-acrylamido - 2-methylpropane sulfonic acid (AMPS) a product of Lubrizol Corp. as a comonomer.

Cationic polyacrylamides (PAMD) were synthesized using 3 mol % of dimethylamino-propyl-methacrylamide (DMAPMA) as the comonomer. This reagent was received as a stabilized liquid from Jefferson Chemical Company. The hydroquinone inhibitor was removed by passing a 50:50 aqueous solution of this reagent through an activated carbon column. The aqueous solution was used immediately after the inhibitor removal stage.

Molecular weight of the respective polymers estimated by measuring the intrinsic viscosity using a capillary viscometer was determined to be 2.2 million for non-ionic polyacrylamide PAM, 2.6 million for anionic polyacrylamide PAMS and 1.9 million for cationic polyacrylamide PAMD.

Sodium dodecylsulfonate: This chemical was purchased from Aldrich Chemical Company and was reported by the manufacturer to be 99.9+% pure. It was used without further purification.

Dodecylamine hydrochloride: This chemical was a product of the Eastman Kodak Company and was used as received.

Inorganic reagents: Fisher certified NaOH and HCl were used for pH modification. ACS Reagent Grade NaCl, a product of Amend Drug and Chemical Company, was used for adjusting the ionic strength.

Water: Triple distilled water of specific conductivity of about 10^{-6} mho from a glass still and collected in a glass container was used in this investigation.

TECHNIQUES

Surface tension was measured by the Wilhelmy Plate Method using a sandblasted platinum plate sensor supported from the arm of a Cahn microbalance (Model 2000). Surface tension and all other measurements, unless otherwise specified, were conducted at room temperature (25°C).

Conductivity was measured using an A. H. Thomas Co., Model 275 conductivity meter.

Relative viscosity measurements were conducted at 30°C using a suspended level Ubbelohde capillary viscometer.

Precipitation studies involved mixing different amount of sulfonate or amine and 4000 mg/kg of the polymer such that the resultant solution contained 1000 mg/kg of the desired polymer and a given concentration of the surfactant. The mixture was shaken overnight in a wrist action shaker and centrifuged at 15,000 rpm for 10 minutes. The supernatant was analyzed for the residual polymer and the surfactant.

The residual amount of C-14 labelled polymers was determined using a Beckman Model LS100C spectrophotometer (liquid scintillation counting technique). Surfactant analysis was conducted using a two phase titration method. It was confirmed that polyacrylamide type polymers did not interfere with the surfactant analysis and vice versa (Moudgil, 1981).

RESULTS AND DISCUSSION

Nonionic Polymer and Surfactant Interaction

Interaction between nonionic polyacrylamide (PAM) and sulfonate or amine is possible through hydrocarbon chain interaction and hydrogen bonding (Arai, Murata, and Shinoda, 1971; Fishman and Eirich, 1971; Jones, 1967; Lewis and Robinson, 1970; Saito, 1967; Schwuger, 1973; Tadros, 1974). Since there is no charge on the polymer, electrostatic type of interaction between PAM and the surfactant is not expected. With chain-chain interaction the polymer is expected to behave like a polyelectrolyte which is characterized by higher relative viscosity than of the polymer alone. Depending upon the extent of interaction between the polymer and the surfactant species, there could be a decrease in the conductivity of the mixture as compared to that when there is no interaction between the two. Surface tension also is exected to be higher when there is hydrocarbon chain interactin between a hydrophilic polymer and an ionic surfactant than in the absence of the polymer. If on the other hand, the associative inter-actions are not significant and the presence of the two species results only in modifications in the solvent power of the medium then instead of an increase in surface tension a decrease may be obtained. The relative viscosity and conductivity under these circumstances may or may not be affected.

PAM-Sulfonate. Surface tension, relative viscosity, conductivity and precipitation results for this system are presented in Figures 1, 2, 3 and 4, respectively. Surface tension of sulfonate in the presence of 1000 mg/kg PAM was determined to be lower up to 10^{-3} kmol/m^3 , than that of the solution containing only sulfonate and higher when the sulfonate concentration was in excess of 10^{-3} kmol/m^3 . The surface tension was observed to attain a constant value at about 7.5×10^{-3} kmol/m^3 sulfonate. The decrease in surface tension below 10^{-3} kmol/m^3 sulfonate in the presence of PAM could have been caused by "salting out" of the sulfonate molecules. The decrease in $\partial\gamma/\partial \log C_T$, on the other hand with increase in sulfonate concentration suggests that the monomer activity in the bulk is not increasing in the presence of PAM, as in the absence of it, indicating some type of association complex formation.

No significant change, however, was observed in the precipitation, relative viscosity and conductivity results suggesting that association complexes if formed in the bulk are not significant enough to affect these measurements.

PAM-Amine. As shown in Figures 2, 3 and 5, no significant change in the relative viscosity, conductivity and precipitation results were observed under the present experimental conditions when polyacrylamide and dodecylamine were mixed. This indicated that, either there was no interaction between the polymer and the amine capable of modifying the polymer molecule conformation or, the meas niques were not sensitive enough to detect the changes caused by si

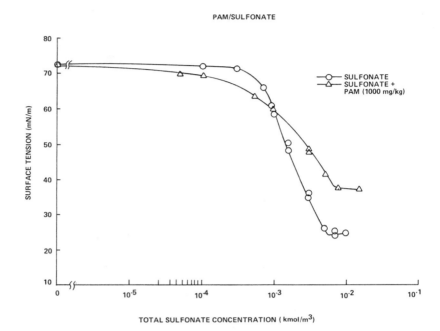

Fig. 1.   Effect of addition of PAM (nonionic polyacrylamide)
on the surface tension of dodecylsulfonate.

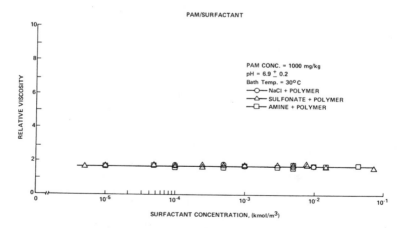

Fig. 2.   Effect of addition of surfactant (dodecylsulfonate/
dodecylamine) on the relative viscosity of PAM (nonionic polyacrylamide).

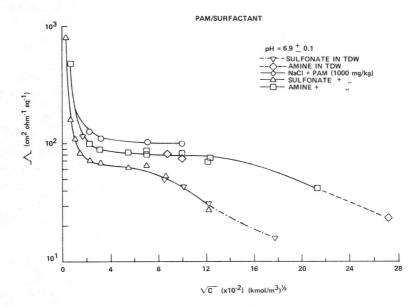

Fig. 3. Effect of addition of PAM (nonionic polyacrylamide) on the conductivity of dodecylsulfonate and dodecylamine.

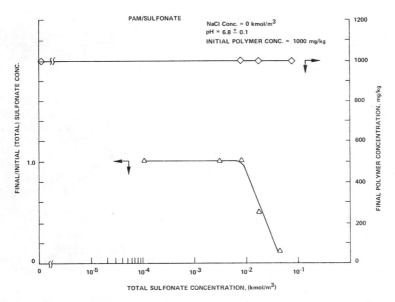

Fig. 4. Effect of dodecylsulfonate concentration on the precipitation behavior of PAM (nonionic polyacrylamide) and dodecylsulfonate mixture.

In summary, associative bulk interactions between nonionic polyacrylamide and both anionic sulfonate and cationic amine under the tested conditions are considered negligible.

Similarly Charged Polymer and Surfactant Interactions

In polymer and surfactant systems where both the species carry similar charges, electrostatic type interactions can be ruled out. The possibility of chain-chain interactions between the polyacrylamides and the surfactants, because of the relatively longer carbon chain of the functional groups attached to the polymers, may be higher than similar interactions with nonionic polyacrylamides. In addition to associative type bulk interactions, changes in different properties may also be caused by modifications in the solvent power of the medium.

PAMS-Sulfonate. The surface tension data presented in Fig. 6 is similar to that obtained for the PAM-Sulfonate system indicating the formation of associative type complexes. The relative viscosity and conductivity results presented in Figs. 7 and 8 respectively, suggest that either there was no associative type bulk interaction between PAMS and sulfonate molecules or the changes caused by such interactions were not significant enough to modify the bulk solution properties.

PAMD-Amine. No change was observed in the conductivity and precipitation behavior as a result of mixing the cationic polyacrylamide and amine indicating that the extent of the bulk complexation, if any, was not significant (See Figs. 9 and 10). The relative viscosity results in the presence of 1000 mg/kg polymer on the hand, were found to be lower up to about 10^{-3} kmol/m^3 amine and to be similar to solution containing polymer and NaCl above it (See Fig. 11). The lowering of the relative viscosity is indicative of the polymer charge neutralization which is not possible in the present case since both the species are similarly charged. The reason for this anamoly is not known at present.

Summarizing the above discussion it is noted that in the case of PAMS and sulfonate and PAMD and amine the two similarly charged polymer-surfactant systems, there was no substantial associative bulk interaction between the different species.

Oppositely Charged Polymer and Surfactant Interactions

In systems where the polymer and the surfactant are oppositely charged, hydrogen bonding and electrostatic, as well as chain-chain, interactions can be expected (Goddard and Hannan, 1976, 1977). Changes in bulk solution properties may also occur through modifications in the solvent power of the medium. Electrostatic interactions should result in lowering of the relative viscosity and conductivity. Complexation between oppositely charged species can, on the other hand, lead also to bulk precipitation.

PAMS-Amine. The PAMS polymer being opositely charged to dodecylamine can be expected to interact electrostatically with the amine molecules. In fact visual observation of a mixture of the two did indicate the formation of a fibrous precipitate which redissolved in the solution upon increasing the amine concentration. This phenomenon of precipitation and redissolution of the complex appears to be similar to that observed in multivalent ions/sulfonate systems (Celik, 1981).

Results obtained for the relative viscosity of amine and PAMS system (Figure 7) showed a minimum with increase in amine concentration and the position of the minimum corresponded to the point of maximum precipitation. However, the concentration at which complete redissolution of the precipitate occurred, was deter-

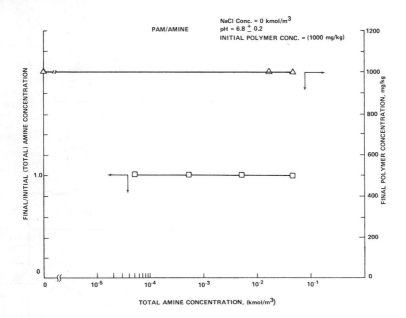

Fig. 5. Effect of dodecylamine concentration on the precipitation behavior of PAM (nonionic polyacrylamide) and dodecylamine mixture.

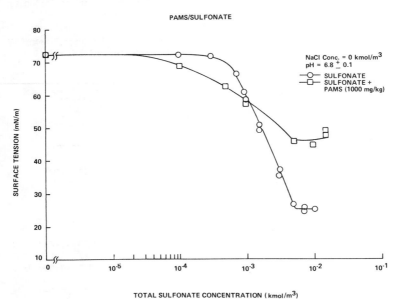

Fig. 6. Effect of addition of PAMS (anionic polyacrylamide) on the surface tension of dodecylsulfonate.

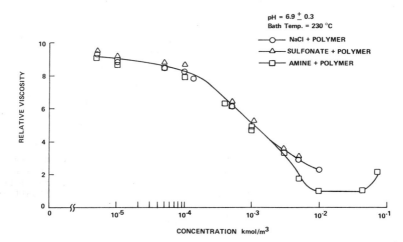

Fig. 7.   Effect of addition of surfactant (dodecylsulfonate/
dodecylamine) on the relative viscosity of PAMS (anionic polyacrylamide).

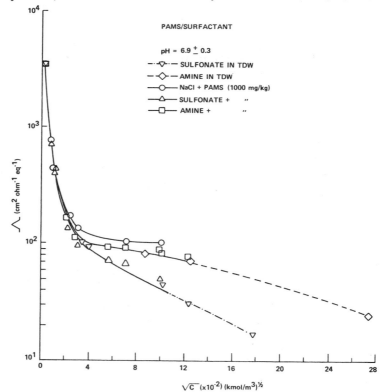

Fig. 8.   Effect of addition of PAMS (anionic polyacrylamide)
on the conductivity of dodecylsulfonate and dodecylamine.

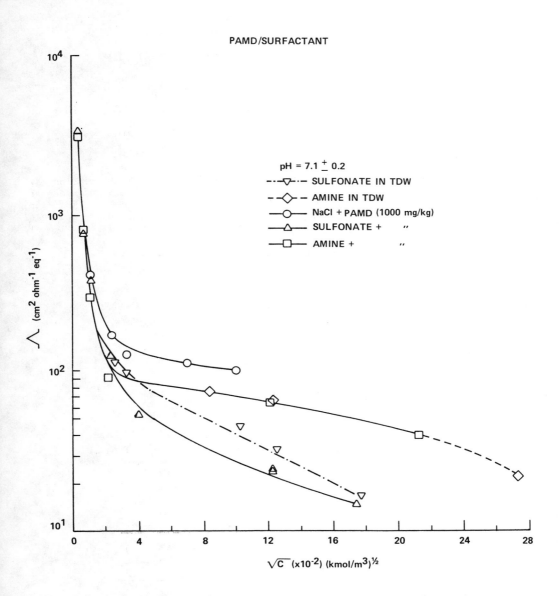

Fig. 9. Effect of addition of PAMD (cationic polyacrylamide) on the conductivity of dodecylsulfonate and dodecylamine.

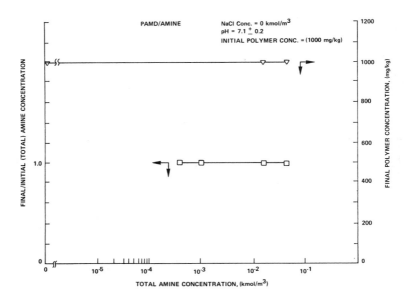

Fig. 10. Effect of dodecylamine concentration on the precipitation behavior of PAMS (cationic polyacrylamide) and
dodecylamine mixture.

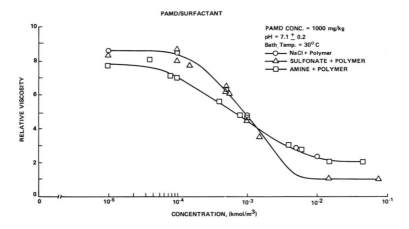

Fig. 11. Effect of addition of surfactant (dodecylsulfonate/
dodecylamine) on the relative viscosity of PAMD (cationic
polyacrylamide).

mined to be lower than that at which the relative viscosity of the mixture started to increase (see Fig. 12). This observation may be partly attributed to the effect of the increase in ionic strength on viscosity. Conductivity measurements are evidently not sensitive under high ionic strength conditions to the changes caused by the precipitation/redissolution phenomena.

Interaction of the anionic polymer with the cationic surfactant leading to the precipitation can be represented by the following reaction

$$P^{-n} + nR^+ \rightleftharpoons PR_n$$

$$K_{sp} = [P^{-n}].[R^+]^n$$

$$\log [P^{-n}] = \log K_{sp} - n \log [R^+]$$

where P^{-n} is the anionic polyacrylamide with n negatively charged sites with which n amine R^+ molecules react to form precipitate PR_n. K_{sp} is the solubility product.

The reaction constant for the precipitation can be obtained by plotting $\log [R^+]$ vs. $\log [P^{-n}]$. In view of the fact that other charged complexes may exist in the aqueous phase, any estimate of the solubility product may only be considered as an apparent value. Assuming that there are no other complexes in the system and only 1:1 interaction between the polymer functional group and the surfactant species occur, the solubility product K_{sp} was estimated to be 3.3×10^{-15}.

Redissolution of the precipitate can take place either due to micellar solubilization or due to complexation in the following manner

$$PR_n + mR^+ = [PR_{n+m}]^{+m}$$

This complexation can result from the chain-chain interaction of an amine molecule already bound to the polymer electrostatically.

PAMD-Sulfonate. The PAMD polymer being oppositely charged to the sulfonate can also be expected to interact electrostatically in the bulk solution. Visual examination of a mixture of the two clearly indicated the formation a fibrous precipitate which, however unlike in the case of PAMS-Amine system did not redissolve into the solution upon increasing the concentration of the sulfonate.

The surface tension behavior of the sulfonate in the presence of the cationic polymer PAMD was, as illustrated in Figure 13, very similar to that obtained with the PAM-sulfonate system, thus suggesting the formation of a complex between the two species.

Results for the conductivity, relative viscosity and precipitation studies (See figures 9, 11 and 14 respectively) provide support for the contention that above 10^{-3} kmol/m^3 sulfonate concentration significant bulk complexation between PAMD and sulfonate molecules is occurring.

The interaction of the sulfonate with the cationic polyacrylamide leading to precipitation can be represented by:

$$P^{+n} + n\bar{R} \rightleftharpoons PR_n$$

$$K_{sp} = [P^{+n}] . [\bar{R}^n]$$

$$\log [P^{+n}] = \log K_{sp} - n \log [R^-]$$

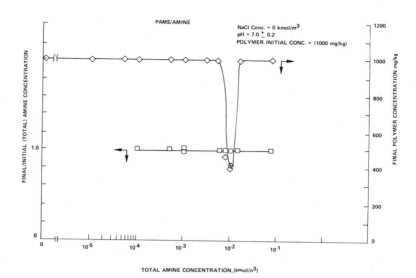

Fig. 12.   Effect of dodecylamine concentration on the precipitation behavior of PAMS (anionic polyacrylamide) and
dodecylamine mixture.

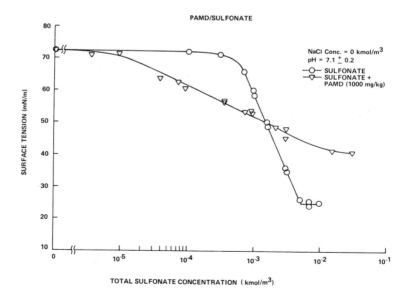

Fig. 13.   Effect of addition of PAMD (cationic polyacrylamide)
on the surface tension of dodecylsulfonate.

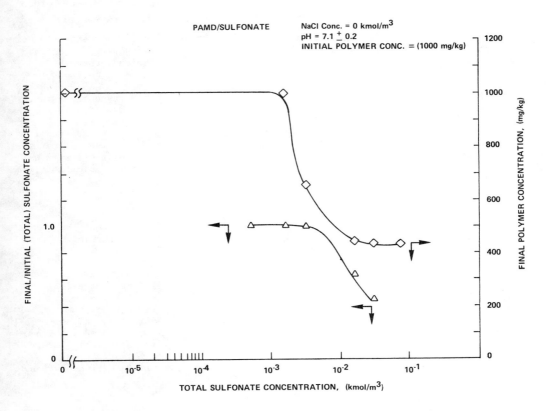

Fig. 14. Effect of dodecylsulfonate concentration on the
precipitation behavior of PAMD (cationic polyacrylamide)
and dodecylsulfonate mixture.

where P^{+n} is the concentration of the charged polyacrylamide which reacts with n molecules of sulfonate R^- to form PR_n precipitate.

The reaction constant, as before, can be obtained by plotting log concentration of the polymer as a function of the log concentration of the sulfonate. Once again, assuming the absence of other complexes and only 1:1 interaction between PAMD functional groups and the sulfonate molecules, K_{sp} was estimated to be 1.6×10^{-14}.

The absence of redissolution may be indicative of the mechanism responsible for solubilizing the precipitate formed. The solubility limit of the sulfonate was determined to be 7.2×10^{-3} kmol/m^3 . If solubilization of the precipitate was occurring through complexation, some evidence of it in terms of higher residual polymer content and possibly higher relative viscosity should have been obbtained, which was not. On the other hand it is conceivable that the CMC of the system was above the solubility limit of the sulfonate and therefore, micellar solubilization did not occur under the present experimental conditions.

CONCLUSIONS

Based on the above discussion it can be concluded that no significant bulk interaction occurs between nonionic polyacrylamide and the anionic dodecylsulfonate or the cationic dodecylaminehydrochloride. On the other hand, in the case of the oppositely charged polymer and surfactant systems bulk complexation takes place to a significant extent. Also, depending on the ionic nature of the polymer and the surfactant they can precipitate and in some cases the precipitate thus formed can redissolve upon increasing the surfacant concentration. Redissolution of the precipitate is suggested to occur either through micellar solubilization and/or complexation. It is to be noted that the type of modification observed in the bulk properties here can conceivably result in major changes in the adsorption behavior of different species at various interfaces and thereby affect processes such as froth flotation and flocculation.

ACKNOWLEDGEMENTS

The authors wish to thank Dr. K. P. Ananthapadmanabhan for helpful discussions and Mrs. Regina Gorelik for help in experimentation. Partial financial support of this investigation by the National Science Foundation (Grant No. DAR-79-09295) and the INCO, Inc. is also acknowledged.

REFERENCES

Arai, H., Murata, M., and Shinoda, K. (1971). The interaction between polymer and surfactant: the composition of the complex between polyvinyl pyrrolidone and sodium alkyl sulfate as revealed by surface tension dialysis and solubilization. J. Colloid Interface Sci., 37, 223-227.
Celik, M. (1981). In annual progress report, "Adsorption of surfactants and polymers on reservoir minerals," by Somasundaran, P., Columbia University, New York.
Fishman, M. L. and Eirich, F. R. (1971). Interactions of aqueous poly (n-vinylpyrrolidone) with sodium dodecyl sulfate I-equilibrium dialysis measurements. J. Phys. Chem. 75 , 3135-3140.
Goddard, E. D., and Hannan, R. B. (1976). Cationic polymer/anionic surfactant interactions. J. Colloid Interface Sci., 55, 73-79.
Goddard, E. D. and Hannan, R. B. (1977). Polymer/surfactant interactions. J. Am. Oil Chemists' Soc., 54, 561-566

Jones, M. (1967). The interaction of sodium dodecyl sulfate with polyethylene oxide. J. Colloid Interface Sci., 23, 36-42.

Lewis, K. E., and Robinson, C. P. (1970). The interaction of sodium dodecyl sulfate with methyl cellulose and polyvinyl alcohol. J. Colloid Interface Sci., 32, 539-546.

Moudgil, B. M. (1981). The Role of Polymer-Surfactant Interactions in Interfacial Processes. Eng. Sc. D. Thesis, Columbia University, New York.

Saito, S. (1967). Solubilization properties of polymer-surfactant complexes. J. Colloid Interface Sci., 24, 227-234.

Schwuger, M. J. (1973). Mechanism of interaction between ionic surfactants and polyglycol ethers in water. J. Colloid Interface Sci., 43, 491-498.

Somasundaran, P. (1969). Adsorption of starch and oleate and interaction between them on calcite in aqueous solutions. J. Colloid Interface Sci., 31, 557-565

Tadros, Th. F. (1974). The interaction of cetyltrimethylammonium bromide and sodium dodecylbenzene sulfonate with polyvinyl alcohol. Adsorption of the polymer-surfactant complexes on silica. J. Colloid Interface Sci., 46, 528-539.

Wada, T., Sekiya, H., and Machi, S. (1975). Radiation induced heterogeneous polymerization of acrylamide in acetone and acetone-water mixtures. J. Appl. Poly. Sci. (Chem.), 13, 2375-2389.

Wada, T., Sekiya, H., and Machi, S. (1976). Synthesis of high molecular weight polyacrylamide flocculant by radiation polymerization. J. Appl. Poly. Sci., 20, 3233-3240.

ELASTOMER/GASOLINE BLENDS INTERACTIONS I.
EFFECTS OF METHANOL/GASOLINE MIXTURES ON ELASTOMERS

by

Ismat A. Abu-Isa
Polymers Department
General Motors Research Laboratories
Warren, MI   48090-9055

ABSTRACT

The effects of methanol/gasoline mixtures on swell properties and tensile properties of selected automotive elastomers were investigated. Two gasolines with aromatic contents of 30% and 50% were used in the investigation. Equilibrium swell measurements and tensile measurements were conducted using ASTM standard procedures. The results show that although few elastomers were affected drastically by pure gasoline (e.g. natural rubber) and a few by methanol (e.g. fluorocarbon elastomer) most of the elastomers were more severely affected by mixtures of the gasoline and methanol rather than the pure components. Presence of higher aromatic content in the methanol/gasoline mixtures led to additional deterioration of properties. The data on all elastomers except the fluorocarbon can be explained in terms of the solubility parameter concept.

Ultimate tensile and ultimate elongation values of elastomer networks were found to be quantitatively related by simple linear equations to the volume fraction of rubber in the swelled networks. They were also found to fit equations derived for equilibrium stress-strain relationships.

KEYWORDS

Elastomer swell; gasoline/methanol blends; solvent swell; tensile properties; stress-strain relationships; solubility parameters.

INTRODUCTION

Methanol and ethanol and methyl-t-butyl ether are blended into some commercial gasolines. It was important to determine the effects of such blends on automotive elastomers especially those used in the fuel system. A summary of the results of our studies was published[1]. The purpose of the current series of publications is to present the detailed data for all elastomers investigated and to provide scientific explanations for the observed effects. This report deals with the effects of methanol/gasoline blends on sixteen automotive elastomers.

166

EXPERIMENTAL

The elastomers investigated in this study are shown in Table I along with their
percent volume swell in Indolene HO-III. Some of these elastomers such as
nitrile, epichlorohydrin and fluorocarbon polymers resist swelling in fuel and
are used in cars as parts for the fuel system. Other elastomers such as neoprene and ethylene-propylene-diene (EPDM) rubber swell to a higher extent in
gasoline and hence cannot be used in the fuel system. However they are used
under the hood and can come in contact with the fuel by an accidental spill or
by malfunction of a diaphragm or valve. The elastomers were formulated using
conventional curing agents, oils, carbon black and other fillers. The formulations of all elastomers tested are shown in Table II. Commercial names and
designations were used in the table in order to enable other investigators to
reproduce the elastomers. Full chemical names of all polymers and ingredients
and names of manufacturers are found in the glossary section of this report.

Two gasolines were used in the investigation. The first is a standard fuel
Indolene HO-III (Amoco), containing 65% by volume paraffins, 30% aromatics and
5% olefins. The second gasoline which we named "spiked" Indolene was prepared
by adding 40 ml of toluene to 100 ml of Indolene HO-III to raise the aromatic
content to 50%. (The concentration of added toluene in spiked Indolene is
28.6%). The paraffin and olefin contents were thus reduced to 46% and 4%
respectively. This gasoline was chosen for investigation because it is close in
its composition to commercial gasolines sold in the U.S. and Europe [2].

Blends of each of the gasolines with methanol were prepared containing 2, 5, 10,
25, 50 and 75% alcohol. The effects of these blends along with the effects of
the pure components on the properties of elastomers were determined.

Three tensile bars and two volume change specimens of each elastomer were
exposed in a test tube to each of the fuels. Exposure for 72 hours at room
temperature was determined to be sufficient for most elastomers to reach equilibrium swell. Tensile properties were determined using the ASTM D412 procedure
for original samples and for fuel exposed samples immediately after removal from
the fuel. Swell measurements were conducted per the ASTM D471 procedure except
that weighing of wet samples was carried out in a closed bottle to prevent fast
loss of fuels by evaporation. To obtain the amount of extractables, the fuel
swelled samples were dried in a vacuum oven at 100°C for 24 hours and then
weighed. In most samples constant weight was achieved within this period of
time.

NMR measurements were also conducted on mixtures of methanol with gasoline to
which either water or ethanol was added in order to explain the high swelling of
fluorocarbon elastomers in methanol. Some results of this investigation will be
discussed later in this report. The experimental conditions and the rest of the
results will be reported elsewhere [3].

RESULTS AND DISCUSSION

I. Effects of the Composition of Gasoline/Methanol Blends on the Volume Swell
and Tensile Properties of Elastomers

In this section we will discuss the effects of the composition of the mixtures
on the properties of elastomers. The composition variables discussed are the
concentrations of methanol and aromatics. The properties of elastomers investigated are the volume swell, amount of extractables, ultimate elongation, ultimate tensile strength and the modulus which is defined as tensile stress at 100%
elongation. Durometer values were measured and are shown in the tables but will

Table I

Elastomers Investigated in Methanol/Gasoline Mixtures

	Elastomer	% Swell in Indolene HO-III
1.	Fluorocarbon	0
2.	Polyether (F-70A)	8
3.	Polyester Urethane	8
4.	Fluorosilicone	14
5.	Epichlorohydrin Copolymer	18
6.	Polysulfide	23
7.	Butadiene-Acrylonitrile	26
8.	Chlorosulfonated Polyethylene	37
9.	Polyacrylate	40
10.	Chlorinated Polyethylene	42
11.	Neoprene	45
12.	Ethylene-Propylene-Diene Terpolymer (EPDM)	122
13.	Natural Rubber	151
14.	Vynathene	157
15.	Styrene-Butadiene Rubber (SBR)	161
16.	Silicone	240

Table II.   Formulation of Investigated Elastomers

Ingredients	Concentration (Phr)	Ingredients	Concentration (Phr)
1.Fluorocarbon		6. Polysulfide	
Viton AHV	100	FA 3000	85.5
Maglite D	15	NW-1	14.5
Diak #1	1.5		
N990 Black	25	FA 3000	
2. Polyether F-70A		Polysulfide FA	100
		N774 Black	60
F-70A	100	Stearic Acid	0.5
N330 Black	60	MBTS (Altax)	0.4
Stearic Acid	4	DPG	0.1
TP759 Plasticizer	10	Protox 166	10
Spider Sulfur	1.2	NA-22	0.1
Curative P	0.9		
		NW-1	
3. Polyester Urethane			
		Neoprene W	100
Vibrathan 5004	100	Maglite D	4
Stearic Acid	0.25	N774 Black	55
Di-Cup 40 KE	4	Stearic Acid	0.5
N330 Black	20	Protox 166	5
4. Fluorosilicone		7. Butadiene-Acrylonitrile	
Silastic LS-70	66.6	Hycar 1042	100
Silastic LS-40	33.3	Zinc Oxide	5
Silastic HT-1	1.0	Sulfur	0.5
Varox	0.75	Methyl Tuads	2
		Santocure	1
5. Epichlorohydrin Copolymer		AgeRite Resin D	2
		N762 Black	70
Herclor C	100		
Zinc Stearate	0.75	8. Chlorosulfonated Polyethylene	
N550 Black	50		
Red Lead	5	Hypalon 48	90
NBC	1	Hypalon 40SS	10
NA-22	1.5	N550 Black	20
		N774 Black	50
		TLD-90	22
		Dioctyl Sebacate	5
		A-C PE 617A	5
		Maglite D	5
		NBC	1
		MBTS	1
		Tetrone A	1
		HVA-2	1

Table II. – Continued

Ingredients	Concentration (Phr)	Ingredients	Concentration (Phr)
9. Polyacrylate		**13. Natural Rubber**	
Hycar 4042	100	Smoked Sheet	100
Stearic Acid	1	Stearic Acid	3
N326 Black	80	Zinc Oxide	3
Spider Sulfur	0.25	Sulfur	3
Potassium Stearate	2.25	Reogen	2
Sodium Stearate	0.75	Pine Tar	1
		Captax	1
10. Chlorinated Polyethylene		AgeRite Resin D	1
		N762 Black	63
CM 0136	100		
Di-Cup 40 KE	6	**14. Vynathene**	
Dythal XL	10		
Paraplex G-62	3	Vynathene EY 904	100
Chlorowax LV	20	N 762 Black	5
TAIC	2	Hisil 215	50
N550 Black	20	AgeRite Resin D	1.5
N990 Black	50	Dioctyl Sebacate	5
		Di-Cup 40 KE	6
11. Neoprene			
		15. Styrene-Butadiene Rubber	
Neoprene W	100		
Stearic Acid	2.5	SBR 1500	100
Zinc Oxide	6	Stearic Acid	2
Maglite D	4.5	Zinc Oxide	3
Sulfur	1	Spider Sulfur	1.4
Dibutyl Sebacate	17	Altax	0.5
AgeRite HP	2	Antozite 67	1
Octamine	2	DPG	0.5
AgeRite DPPD	2	N220 Black	50
N990 Black	15		
N550 Black	40	**16. Silicone**	
12. Ethylene-Propylene-Diene Terpolymer		Silicone GP-70	100
		HT-1	1
		Varox	1
Vistalon 6505	100		
Zinc Oxide	5		
Di-Cup 40 KE	10		
AgeRite Resin D	0.25		
Sunpar 2280	60		
TMPT	1		
N762 Black	70		
N550 Black	70		

not be discussed because of the inexact nature of this measurement. For convenience this part of the discussion will be divided into three sections dealing
with fuel resistant elastomers, medium fuel resistant elastomers and poor fuel
resistant elastomers.

A. Fuel Resistant Elastomers

These are arbitrarily defined as those elastomers whose equilibrium swell in
Indolene HO-III is less than 30% by volume. They include the following
elastomers: fluorocarbon, polyether, polyester urethane, fluorosilicone,
epichlorohydrin copolymer, polysulfide and butadiene-acrylonitrile (nitrile).
The complete results on the effects of the various methanol/Indolene blends on
these elastomers are shown in Table III. A typical example of the influence of
methanol concentration and aromatic content on this type of elastomer is found
in Figure 1. The figure shows the variation in volume swell, tensile strength
and elongation of nitrile elastomer with the composition of the fuel. The
volume swell in Indolene HO-III is 26%. Addition of methanol into Indolene
increases the swell of nitrile elastomer to a maximum value of 53% occurring in
a mixture containing about 15% methanol. As the concentration of methanol in
the mixture is increased swell decreases systematically to a value of 14% in
pure methanol. Increase in the swell of the elastomer leads to decrease in
ultimate tensile strength and ultimate elongation as seen in Figure 1. Minimum
values in tensile strength and elongation are thus observed after exposure to
blends where maximum swell occurs. This dependence of tensile strength and
elongation on volume swell was observed for all elastomers and will be quantitatively examined in a later section of this report.

The concentration of methanol in the fuel mixture at which maximum swell occurs
varies with the type of elastomer. For polyether, polysulfide, polyester
urethane and nitrile elastomers maximum swell occurs in a mixture containing 10
to 20% methanol (Table III). For fluorosilicone and polyepichlorohydrin copolymer maximum swell occurs at higher concentrations of methanol ranging between 25
and 45%. For fluorocarbon elastomer maximum swell occurs in pure methanol. The
above observations can be explained on the basis of the solubility parameters of
the elastomers and the fuel mixtures in all cases except the fluorocarbon
elastomer. Instead of having a maximum swell, as predicted by the solubility
parameter, in a mixture containing about 15% methanol, swell continues to
increase reaching a high value of 100% in pure methanol. A systematic decrease
in ultimate tensile strength and ultimate elongation is observed with increase
in methanol concentration as seen in Figure 2.

Increasing the aromatic content of the fuel mixtures increases the volume swell
of the elastomers as seen in Figure 1 and Table IV. Swell values in spiked
Indolene containing 50% aromatics are consistently higher than in Indolene
HO-III with 30% aromatic content. The aromatic content was raised to 50% in
spiked Indolene by adding 40 ml of toluene to 100 ml of Indolene HO-III.

Comparing the volume swell results of spiked Indolene with the Indolene HO-III
mixtures containing 10% methanol (Table IV), it is seen that the addition of the
10% alcohol is generally more detrimental, i.e. leads to higher swell and lower
tensile properties, than the addition of 28.6% toluene (compare columns 4 and 5
and columns 10 and 11 of Table IV). The simultaneous addition of toluene and
methanol to Indolene has an additive effect rather than a synergistic effect as
seen by comparisons of volume swell results and ultimate tensile properties of
the elastomers after exposure to spiked Indolene containing 10% methanol with
the rest of the data in Table IV. For example adding 10% methanol to Indolene
HO-III increases the volume swell of polysulfide from 23% to 31% i.e. by 8%.
Addition of 28.6% toluene increases the volume swell by 12% (from 23 to 35%),
while addition of both results in a 20% (from 23% to 43%) volume increase.

Table III.

Effect of Methanol/Gasoline Mixtures on
Properties of Fuel Resistant Elastomers After
Immersion for 72 Hours at Room Temperature

Elastomer	Gasoline	% Methanol	Tensile Strength MPa	Elongation %	Modulus at 100% Elongation MPa	Durometer Shore A Points	Volume Change %	Extractables %
Fluorocarbon	–	–	16.8	200	5.7	73	–	–
	Indolene	0	15.8	232	5.3	76	0	0
		2	12.1	199	4.5	68	2	0
		5	11.7	219	4.0	63	11	0
		10	8.7	178	3.7	60	27	0
		25	5.3	127	3.9	59	46	0
		50	5.6	116	4.6	58	63	0
		75	4.8	96	–	57	95	0
		100	4.3	87	–	57	100	0
	Spiked Indolene	0	15.1	208	5.4	74	1	0
		2	11.6	197	4.5	69	6	0
		5	11.2	197	4.4	68	11	0
		10	10.0	162	4.0	60	21	0
		25	6.4	141	4.0	58	37	0
		50	5.6	118	4.6	51	61	0
		75	4.6	93	–	52	91	0
Polyether F-70A	–	–	9.6	203	4.4	78	–	–
	Indolene	0	7.1	159	4.1	70	8	6
		2	5.8	158	3.2	64	18	6
		5	5.2	127	3.6	64	24	6
		10	4.6	112	3.8	60	29	7
		25	4.6	116	3.8	63	29	7
		50	4.8	111	4.1	62	23	8
		75	4.6	114	4.0	68	19	7
		100	6.5	128	4.4	73	11	7
	Spiked Indolene	0	4.8	98	–	69	22	7
		2	4.5	113	3.9	64	30	7
		5	3.1	94	–	57	38	7
		10	3.8	86	–	60	45	8
		25	3.6	81	–	60	48	9
		50	3.9	94	–	63	40	9
		75	4.6	115	3.8	61	25	8
Polyester Urethane	–	–	27.9	580	3.0	65	–	–
	Indolene	0	21.0	495	2.9	62	8	1
		2	15.1	384	2.5	55	24	2
		5	12.2	339	2.2	46	37	2
		10	9.8	232	2.0	45	42	2
		25	11.1	323	2.3	50	39	2
		50	13.6	362	2.3	51	35	2
		75	13.4	356	2.4	55	27	2
		100	15.8	396	2.6	51	18	2
	Spiked Indolene	0	16.9	409	2.9	55	21	2
		2	12.9	350	2.5	55	34	2
		5	10.2	304	2.2	48	50	2
		10	8.1	255	2.2	41	58	2
		25	9.2	342	2.0	47	59	3
		50	11.2	316	2.1	47	52	3
		75	9.9	297	2.3	44	39	3
Fluorosilicone	–	–	7.5	241	2.8	62	–	–
	Indolene	0	6.4	246	2.1	–	14	2
		2	4.3	192	1.8	–	16	2
		5	4.2	185	1.9	–	19	3
		10	3.8	160	2.1	–	21	3
		25	3.7	169	1.9	–	22	3
		50	3.8	179	1.7	–	21	4
		75	4.6	196	1.9	–	15	4
		100	4.5	246	1.8	–	8	3
	Spiked Indolene	0	5.2	202	2.2	–	18	2
		2	3.7	181	2.0	–	19	3
		5	3.0	172	1.9	–	21	3
		10	4.0	176	1.9	–	22	3
		25	3.7	168	1.9	–	23	3
		50	3.7	164	2.0	–	22	3
		75	4.2	185	2.0	–	15	4
Epichlorohydrin Copolymer	–	–	15.3	329	5.3	76	–	–
	Indolene	0	13.3	257	5.2	71	18	1
		2	11.4	223	4.7	65	29	1
		5	11.0	215	5.0	63	46	1
		10	10.0	200	4.5	58	56	1
		25	9.5	185	4.7	60	62	2
		50	10.3	200	4.8	61	60	2
		75	11.3	212	5.0	64	50	2
		100	12.1	229	5.1	65	31	2
	Spiked Indolene	0	11.4	207	5.4	65	32	1
		2	11.0	214	5.0	60	44	1
		5	8.9	151	5.8	59	63	2
		10	8.4	148	5.4	57	77	2
		25	8.8	148	5.7	59	93	3
		50	9.1	161	5.3	60	94	3
		75	10.3	180	5.3	58	61	3

Table III. - Continued

Elastomer	Gasoline	% Methanol	Tensile Strength MPa	Elongation %	Modulus at 100% Elongation MPa	Durometer Shore A Points	Volume Change %	Extractables %
Polysulfide	-	-	10.6	267	5.1	-	-	-
	Indolene	0	8.1	205	4.3	62	23	2
		2	7.7	205	4.0	62	28	2
		5	6.5	173	3.8	62	30	2
		10	6.4	185	3.5	59	31	2
		25	6.8	190	3.6	58	30	2
		50	6.8	168	4.0	60	28	2
		75	7.6	198	4.0	63	23	2
		100	8.8	220	4.4	64	18	2
	Spiked Indolene	0	6.2	193	2.2	53	35	2
		2	5.6	163	2.4	50	42	2
		5	5.3	168	2.6	51	42	3
		10	4.9	155	2.6	42	43	3
		25	5.1	153	2.6	49	40	3
		50	5.8	167	2.2	58	33	2
		75	7.3	200	1.7	60	26	2
Butadiene Acrylonitrile (Nitrile)	-	-	17.2	256	6.7	75	-	-
	Indolene	0	10.0	163	5.8	61	26	1
		2	8.3	137	5.8	61	36	2
		5	6.9	113	6.0	56	47	3
		10	7.5	111	6.7	58	53	2
		25	7.0	103	6.8	60	52	3
		50	7.9	113	6.9	59	46	4
		75	9.0	135	6.7	58	34	3
		100	11.0	190	5.5	65	14	2
	Spiked Indolene	0	7.5	116	6.4	60	51	2
		2	6.3	105	6.0	55	61	3
		5	5.4	85	-	57	72	3
		10	5.5	87	-	53	81	4
		25	5.4	83	-	53	82	4
		50	6.0	99	-	51	68	4
		75	7.8	144	5.0	58	40	4

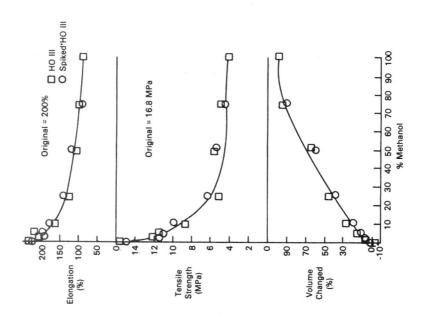

Figure 2. Tensile Elongation, Tensile Strength and Volume Change of Fluorocarbon Elastomer After Exposure to Methanol/ Gasoline Compositions.

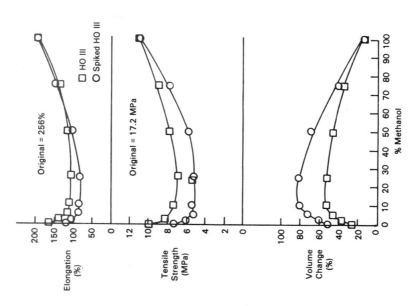

Figure 1. Tensile Elongation, Tensile Strength and Volume Change of Nitrile Elastomer After Exposure to Methanol/Gasoline Compositions.

Table IV.

Volume Swell and Tensile Strength of Fuel Resistant Elastomers After Exposure to Methanol, Indolene HO-III, Spiked Indolene, and 10% Methanol Mixtures with Each of the Indolenes

| Elastomer | Volume Swell (%) After 72 hr. Immersion in | | | Mixtures of 10% Methanol in | | Original Tensile Strength MPa | Tensile Strength (MPa) After 72 hr. Immersion in | | | Mixtures of 10% Methanol in | |
	Methanol	Indolene	Spiked Indolene	Indolene	Spiked Indolene		Methanol	Indolene	Spiked Indolene	Indolene	Spiked Indolene
Fluorocarbon	100	0	1	27	21	16.8	4.3	15.8	15.1	8.7	10
Polyether F-70A	11	8	22	29	45	9.6	6.5	7.1	4.8	4.6	3.8
Polyester Urethane	18	8	21	42	58	27.9	15.8	21.0	16.9	9.8	8.1
Fluorosilicone	8	14	18	21	22	7.5	4.5	6.4	5.2	3.8	4.0
Epichlorohydrin Copolymer	31	18	32	56	77	15.3	12.1	13.3	11.4	10.0	8.4
Polysulfide	18	23	35	31	43	10.6	8.8	8.1	6.2	6.4	4.9
Butadiene-Acrylonitrile	14	26	51	53	71	17.2	11.0	10.0	7.5	7.5	5.5

Quantitative relationships between volume swell values and modulus values of the different elastomers shown in Table III will be examined later. In general lower modulus values are observed at higher swell. Relationships between the amount of extractables and degree of swell can only be qualitatively described. In general a fuel mixture that swells the elastomer to a higher degree also extracts a larger fraction of the elastomer (Table III).

B. Medium Fuel Resistant Elastomers

These are again arbitrarily defined as those elastomers whose equilibrium swell in Indolene HO-III at room temperature falls between 30% and 100%. They include chlorosulfonated polyethylene, polyacrylate, chlorinated polyethylene and neoprene. The detailed results on the effects of composition of the methanol/gasoline mixtures on volume swell, tensile strength, ultimate elongation, modulus (stress at 100% elongation), durometer hardness, and the amount of extractables of these elastomers are shown in Table V.

Three of the elastomers, namely, chlorosulfonated polyethylene, chlorinated polyethylene and neoprene are only slightly adversely affected by the presence of alcohol in gasoline and only at low concentration of alcohol (2% to 10%). Figure 3 shows the variation in volume swell, tensile strength, and ultimate elongation of chlorinated polyethylene as a function of methanol concentration. The volume swell increases from a value of 40% for samples exposed to Indolene HO-III to a maximum swell value of 46% for samples exposed to a mixture of Indolene HO-III containing 2% methanol. Swell decreases thereafter with increased methanol concentration to a value of -2% after exposure to pure methanol. Tensile strength values of fuel exposed samples show systematic increase with the increase in methanol concentration of the fuel, whereas ultimate elongation values exhibit a shallow minimum at about 5% methanol concentration followed by steadily increasing values with increased methanol concentration (Figure 3).

The polyacrylate elastomer has a different behavior from that of the above three chlorinated elastomers. Its volume swell in pure methanol (94%) is quite higher than the volume swell in Indolene HO-III (40%) and equals the swell in spiked Indolene (95%) as seen in Figure 4. Even higher volume swell values of this elastomer are obtained after exposure to methanol/Indolene mixtures, the maximum being 165% in a mixture containing 35% methanol. Degradation of tensile strength and ultimate elongation is again observed with increased swell.

The influence of higher aromatic content on properties of these elastomers can be seen in Figures 3 and 4 and Table VI. In the case of the three chlorinated elastomers increasing the aromatic content has a much greater detrimental effect than the addition of alcohol. The reverse effect is observed in the case of the polyacrylate. The chemical composition of the polymers and hence the difference in solubility parameters explains the effects of alcohol and aromatic content. Qualitatively, polyacrylate being more polar interacts better with alcohol than with aromatic hydrocarbons. As observed in the case of the fuel resistant elastomers the incorporation of both aromatics and alcohol into gasoline has an additive rather than a synergistic effect (Table VI).

It is interesting to note that except for the polyacrylate the resistance of these medium fuel resistant elastomers to 10% methanol/90% Indolene mixture is better than that of the more highly fuel resistant nitrile elastomer. From Table IV the volume swell of nitrile in this mixture is 53% as compared to the values reported in Table VI of 41% for chlorosulfonated polyethylene, 42% for chlorinated polyethylene and 44% for neoprene.

Table V.

Effects of Methanol/Gasoline Mixtures on Properties of Medium Fuel
Resistant Elastomers After Immersion for 72 Hours at Room Temperature

Elastomer	Gasoline	% Methanol	Tensile Strength MPa	Elongation %	Modulus at 100% Elongation MPa	Durometer Shore A Points	Volume Change %	Extractables %
Chlorosulfonated Polyethylene	-	-	20.7	105	20.0	94	-	-
	Indolene	0	12.2	78	-	-	37	0
		2	10.9	79	-	-	41	1
		5	10.8	85	-	-	40	1
		10	9.1	73	-	-	41	1
		25	12.2	98	-	-	38	1
		50	12.9	100	12.9	-	31	0
		75	16.3	110	14.9	-	12	0
		100	19.1	105	18.3	-	1	0
	Spiked Indolene	0	5.8	44	-	-	61	2
		2	7.3	61	-	-	68	2
		5	8.0	66	-	-	68	2
		10	6.5	58	-	-	66	2
		25	7.5	69	-	-	55	2
		50	10.5	85	-	-	41	1
		75	12.9	112	11.4	-	18	0
Polyacrylate	-	-	9.6	281	4.1	76	-	-
	Indolene	0	4.1	130	3.1	45	40	0
		2	3.0	95	-	48	52	0
		5	2.1	68	-	47	82	0
		10	1.8	63	-	45	112	0
		25	1.6	64	-	44	155	0
		50	1.5	56	-	43	153	1
		75	1.7	58	-	43	133	1
		100	1.9	81	-	32	94	0
	Spiked Indolene	0	2.1	70	-	50	95	0
		2	1.8	62	-	40	122	0
		5	1.7	62	-	46	135	1
		10	1.4	46	-	44	162	0
		25	1.4	52	-	45	193	1
		50	1.5	53	-	43	181	0
		75	1.6	60	-	45	142	1
Chlorinated Polyethylene	-	-	15.6	295	4.2	67	-	-
	Indolene	0	9.5	207	3.3	45	42	10
		2	9.1	193	3.5	40	46	9
		5	9.1	191	3.5	49	43	10
		10	9.0	200	3.1	44	42	11
		25	9.6	220	3.2	46	37	10
		50	10.6	236	2.9	50	31	9
		75	12.2	285	2.9	52	14	7
		100	13.5	306	3.4	70	-2	2
	Spiked Indolene	0	6.4	140	3.7	42	83	11
		2	6.5	135	4.3	45	87	11
		5	6.6	137	4.2	44	84	1
		10	6.9	138	4.3	45	87	11
		25	7.5	157	3.8	48	60	11
		50	8.9	194	3.3	50	38	9
		75	11.2	240	3.0	55	15	7
Neoprene	-	-	15.8	318	3.3	58	-	-
	Indolene	0	9.6	205	3.3	-	45	11
		2	9.3	201	3.2	-	46	13
		5	9.6	204	3.3	-	45	13
		10	9.4	203	3.2	-	44	13
		25	9.8	215	3.0	-	43	13
		50	10.8	214	3.2	-	38	13
		75	11.9	251	3.0	-	23	11
		100	15.3	322	3.1	-	-4	8
	Spiked Indolene	0	6.0	132	4.0	-	96	12
		2	6.7	145	3.9	-	94	13
		5	7.3	156	3.7	-	86	13
		10	8.1	170	3.7	-	80	14
		25	8.0	177	3.3	-	63	14
		50	9.2	205	3.1	-	43	13
		75	11.3	247	2.9	-	24	11

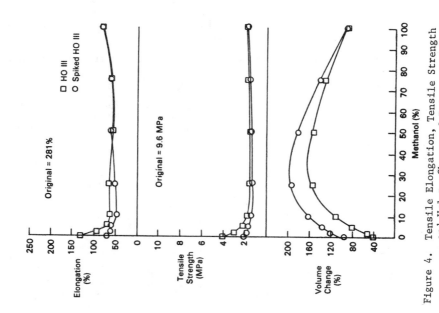

Figure 4.   Tensile Elongation, Tensile Strength and Volume Change of Polyacrylate Elastomer After Exposure to Methanol/Gasoline Compositions.

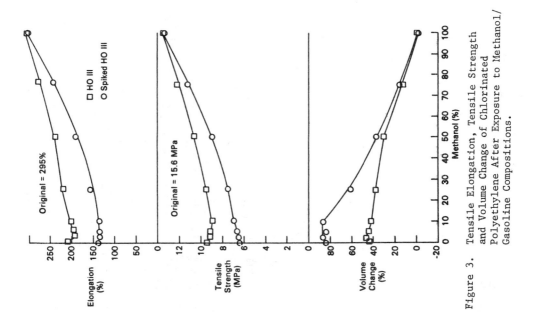

Figure 3.   Tensile Elongation, Tensile Strength and Volume Change of Chlorinated Polyethylene After Exposure to Methanol/Gasoline Compositions.

Table VI.

Volume Swell and Tensile Strength of Medium Fuel Resistant Elastomers After Exposure to Methanol, Indolene HO-III, Spiked Indolene, and 10% Methanol Mixtures with Each of the Indolenes

| Elastomer | Volume Swell (%) After 72 hr. Immersion in | | | Mixtures of 10% Methanol in | | Original Tensile Strength MPa | Tensile Strength (MPa) After 72 hr. Immersion in | | | Mixtures of 10% Methanol in | |
	Methanol	Indolene	Spiked Indolene	Indolene	Spiked Indolene		Methanol	Indolene	Spiked Indolene	Indolene	Spiked Indolene
Chlorosulfonated Polyethylene	1	37	61	41	66	20.7	19.1	12.2	5.8	9.1	6.5
Polyacrylate	94	40	95	112	162	9.6	1.9	4.1	2.1	1.8	1.4
Chlorinated Polyethylene	-2	42	83	42	87	15.6	13.5	9.5	6.4	9.0	6.9
Neoprene	-4	45	96	44	80	15.8	15.3	9.6	6.0	9.4	8.1

C. Poor Fuel Resistant Elastomers

These are defined as elastomers whose equilibrium volume swell in Indolene at
room temperature exceeds 100%. They include the following elastomers, ethylene-
propylene-diene terpolymer (EPDM), natural rubber, vinyl acetate-ethylene
copolymer (Vynathene), styrene-butadiene (SBR) and silicone. All these
elastomers except Vynathene exhibit low volume swell in methanol. The effects
of fuel composition on volume swell, tensile strength, ultimate elongation,
modulus, hardness and amount of extractables are shown in Table VII. Plots of
the effects of aromatic and methanol contents on selected properties of natural
rubber are also shown in Figure 5. From this data it is obvious that for all
polymers in this category except Vynathene addition of methanol leads to only a
slight increase in volume swell at low concentrations (2%) followed by large
systematic decrease as the concentration of methanol is increased. In the case
of Vynathene very large increases in volume swell are observed with the addition
of methanol up to a concentration of 25%, followed by a systematic decrease at
higher methanol concentrations.

The effect of higher aromatic content can be compared with the effect of
methanol addition from data shown in Table VIII. Except for Vynathene, fuels
containing higher aromatic content increase the swell of the elastomers in this
category only slightly, while the addition of 10% alcohol generally leads to
decreased swell. In the case of Vynathene both aromatics and methanol increase
the swell appreciably. The effects of addition of both toluene and methanol is
again additive in the case of these poor fuel resistant elastomers.

II. Solubility Parameter

The solubility parameter is a basic property of all materials. It is defined as
the square root of cohesive energy density or energy of vaporization per unit
volume.

$$\delta = (C.E.D.)^{1/2} = \left[\frac{\Delta H - RT}{V_m}\right]^{1/2}$$

In the above equation C.E.D. is the cohesive energy density, ΔH is enthalpy of
vaporization, V_m is the molar volume, R is the gas constant and T is the absolute temperature. The cohesive energy between molecules of a material can be
considered to be the result of three types of interaction [4]. The first type
is the London dispersion forces which arise from fluctuating atomic dipoles
caused by the presence of a positive nucleus with electrons rotating around
it. The energy contribution of this interaction to the solubility parameter is
designated as δ_d. The second type is polar interaction between molecules caused
by dipole-dipole or dipole-induced dipole forces. The contribution of this to δ
is designated as δ_p. The third interaction is hydrogen bonding which is a
special type of polar interaction occuring between highly electronegative ele-
ments such as oxygen, fluorine, nitrogen or phosphorous and polarized hydrogen
atoms. The solubility parameter contribution of this interaction is designated
as δ_H. The total solubility parameter of materials in terms of these forces is:

$$\delta = \sqrt{\delta_d^2 + \delta_p^2 + \delta_H^2}$$

The solubility parameter of liquids can be calculated from cohesive energy
density values. Assignment to dispersion, polar and hydrogen bonding can be
made through calculations from thermodynamic properties of the solvents [5].
The solubility parameters of polymers can be calculated from measured heats of
vaporization or estimated from the contributions by functional groups in the
polymer [6]. It can also be indirectly calculated from surface tension or solubility data in different solvents [7].

Table VII.

Effects of Methanol/Gasoline Mixtures on Properties of Poor Fuel
Resistant Elastomers After Immersion for 72 Hours at Room Temperature

Elastomer	Gasoline	% Methanol	Tensile Strength MPa	Elongation %	Modulus at 100% Elongation MPa	Durometer Shore A Points	Volume Change %	Extractables %
Ethylene-Propylene-Diene Terpolymer	-	-	12.4	236	4.1	65	-	-
	Indolene	0	4.7	87	-	46	122	16
		2	4.9	90	-	47	125	16
		5	4.5	92	-	45	119	16
		10	4.3	86	-	45	109	16
		25	4.6	92	-	48	103	14
		50	5.0	93	-	45	101	7
		75	7.1	146	4.2	50	48	0
		100	12.8	262	3.8	70	0	0
	Spiked Indolene	0	4.3	89	-	45	134	16
		2	3.9	83	-	48	134	17
		5	3.9	85	-	48	122	16
		10	4.9	97	-	41	106	16
		25	4.3	92	-	45	89	14
		50	4.5	95	-	47	83	4
		75	7.2	161	3.9	50	35	0
Natural Rubber	-	-	17.3	449	3.1	64	-	-
	Indolene	0	4.9	150	3.4	44	151	2
		2	4.1	144	2.9	40	163	3
		5	4.6	158	3.0	43	157	3
		10	4.7	158	3.1	43	148	4
		25	5.2	187	2.9	41	135	4
		50	7.0	279	2.5	41	103	3
		75	12.1	456	2.2	54	45	2
		100	16.3	505	2.6	60	1	1
	Spiked Indolene	0	4.3	151	2.9	41	179	3
		2	3.5	130	2.6	40	194	3
		5	3.5	125	2.8	39	183	4
		10	3.9	147	2.6	40	173	4
		25	4.1	160	2.7	40	137	4
		50	6.6	272	2.7	45	85	3
		75	11.7	389	2.4	50	34	2
Vynathene	-	-	18.9	547	2.5	70	-	-
	Indolene	0	3.3	177	1.7	35	157	4
		2	2.0	142	1.2	28	177	4
		5	1.9	123	1.4	22	205	4
		10	1.6	92	-	24	271	4
		25	1.3	72	-	26	313	4
		50	2.0	108	1.7	26	246	4
		75	3.0	195	1.1	27	111	4
		100	10.9	517	0.9	35	29	4
	Spiked Indolene	0	1.9	96	-	25	285	4
		2	1.7	85	-	21	311	4
		5	1.4	76	-	25	340	5
		10	1.4	71	-	20	378	5
		25	1.6	78	-	25	390	5
		50	2.0	103	1.9	24	246	4
		75	3.3	209	1.0	25	111	4
Styrene-Butadiene Rubber	-	-	24.5	462	2.9	66	-	-
	Indolene	0	3.4	98	3.4	35	161	4
		2	3.6	102	3.2	36	165	5
		5	3.6	97	-	37	161	5
		10	3.4	99	-	35	154	5
		25	3.8	119	2.9	34	138	5
		50	4.0	117	3.0	34	112	3
		75	17.5	358	2.5	60	85	3
		100	23.5	455	2.7	60	0	1
	Spiked Indolene	0	3.1	77	-	33	220	5
		2	2.9	73	-	35	220	5
		5	2.8	77	-	32	209	6
		10	3.0	82	-	39	195	6
		25	3.0	92	-	30	152	6
		50	4.2	133	2.5	36	87	4
		75	7.8	208	2.2	41	37	3
Silicone	-	-	6.1	337	1.69	67	-	-
	Indolene	0	3.0	82	-	25	240	1
		2	2.6	79	-	28	250	1
		5	2.1	78	-	28	254	1
		10	2.1	87	-	20	243	2
		25	2.2	92	-	25	218	3
		50	2.8	120	2.1	30	123	4
		75	7.0	264	2.1	43	37	7
		100	9.8	328	2.8	61	1	4
	Spiked Indolene	0	2.4	78	-	31	248	2
		2	1.7	63	-	22	252	2
		5	2.4	85	-	24	251	2
		10	2.2	87	-	23	244	2
		25	2.2	102	2.1	21	196	4
		50	3.0	142	1.8	28	86	8
		75	6.7	280	2.1	45	28	8

Table VIII.

Volume Swell and Tensile strength of Poor Fuel Resistant Elastomers After Exposure to Methanol, Indolene HO-III, Spiked Indolene and 10% Methanol Mixtures With Each of the Indolenes

| Elastomer | Volume Swell (%) After 72 hr. Immersion in | | | Mixtures of 10% Methanol in | | Original Tensile Strength MPa | Tensile Strength (MPa) After 72 hr. Immersion in | | | Mixtures of 10% Methanol in | |
	Methanol	Indolene	Spiked Indolene	Indolene	Spiked Indolene		Methanol	Indolene	Spiked Indolene	Indolene	Spiked Indolene
EPDM	0	122	134	109	106	12.4	12.8	4.7	4.3	4.3	4.9
Natural Rubber	1	151	179	148	173	17.3	16.3	4.9	4.3	4.7	3.9
Vynathene	29	157	285	271	378	18.9	10.9	3.3	1.9	1.6	1.4
SBR	0	161	220	154	195	24.5	23.5	3.4	3.1	3.4	3.0
Silicone	1	240	248	243	244	6.1	9.8	3.0	2.4	2.1	2.2

For polymer–solvent pairs the optimum solubility occurs when $\Delta H_{mixing} = 0$.
ΔH_{mixing} is related to the solubility parameters of the solvent and the polymer
as follows [7]:

$$\Delta H_{mixing} = V_m \, (\delta_o - \delta_p)^2 \, \phi_o \, \phi_p$$

where V_m is the molar volume of the solvent, δ_o and δ_p are the solubility
parameter of the solvent and the polymer respectively, ϕ_o and ϕ_p are the corresponding volume fractions. For ΔH_{mixing} to become equal to zero, the solubility
parameter of the polymer should be equal to that of the solvent, i.e., $\delta_p = \delta_o$.
In the case of crosslinked elastomers the polymer cannot be dissolved because of
the three dimensional network structure. Instead swelling is observed. The
degree of swelling has been investigated by Flory and Rehner [8] and later by
Flory [9] and by Langley and Ferry [10] and found to be a function of the degree
of crosslinking or the molecular weight between crosslinks M_c and the polymer–
solvent interaction parameter μ. Higher M_c or lower μ leads to increased
swelling [7]. μ is related to the solubility parameters in the following way.

$$\mu = \mu_s + \frac{V_m}{RT} \, (\delta_o - \delta_p)^2$$

Hence at constant degree of crosslinking the optimum swelling of an elastomer in
a solvent occurs when $\delta_o = \delta_p$. In the above equation μ_s is a constant.

In order to find out whether or not the volume swell data obtained for the
various polymers and shown in Tables III, V and VII can be explained in terms of
solubility parameters one needs to compare established solubility parameters of
polymers with those calculatd from maximum swell data. The solubility parameters of fourteen of the elastomers investigated have been reported in the
literature [11–14] and are shown in column 2 of Table IX. The next column of
the table is generated from the present data and lists the concentrations of
methanol in methanol/Indolene mixtures required for maximum swell of each of the
polymers. The solubility parameter of these mixtures can be calculated assuming
that the total solubility parameter is a weighted average of the solubility
parameter of the components. The solubility parameter of methanol [11] is 14.5.
The solubility parameter of Indolene is 7.9 as calculated from the fuel composition and from solubility parameters [11] of the different components of the
fuel.

From the above, the solubility parameters of the methanol/Indolene mixtures in
which maximum swell occurs are calculated and listed in the last column of
Table IX. They are also plotted versus solubility parameters of the elastomers
as shown in Figure 6. For all elastomers except fluorocarbon there is a good
agreement with the solubility parameter concept which predicts $\delta_p = \delta_o$ at
maximum swell. The apparent discrepancy in the case of the fluorocarbon elastomer will be discussed in the next section.

III. High Swell of Fluorocarbon Elastomer in Methanol

The fluorocarbon used in this investigation is Viton A which is a copolymer of
vinylidene fluoride and hexafluoropropylene [15]. The overall solubility
parameter of the polymer is reported [12] to be 8.7 $(cal/cc)^{1/2}$. The
contributions of nonpolar, polar, and hydrogen bonding forces to the overall
solubility parameter are $\delta_d = 7.5$, $\delta_p = 3.5$ and $\delta_H = 2.6$ $(cal/cc)^{1/2}$. From the
solubility parameter values of the methanol/Indolene mixtures, this fluorocarbon
elastomer should show maximum swell in a mixture containing 15% methanol. The
actual swell data in Figure 2 shows that the fluorocarbon elastomer does not
exhibit a maximum swell at 15% methanol but instead swelling continues to
increase reaching a high value of 100% in pure methanol.

Table IX.

Comparison of Solubility Parameters $(cal/cc)^{1/2}$ of Elastomers with
Solubility Parameters of Methanol/Gasoline Mixtures at Maximum Swell

Solubility Parameter of Elastomer Elastomer	% Methanol in Indolene for Maximum Swell (from Literature)	Solubility Parameter of Methanol/Indolene Mixture at Maximum Swell
Fluorocarbon 6.6 - 8.7	100	14.5
Polyester Urethane	9.8 - 10.3	20	9.2
Fluorosilicone	8.8	16	9.0
Epichlorohydrin Copolymer	10.5 - 11.0	40	10.5
Butadiene-Acrylonitrile	9.2 - 9.9	15	8.9
Chlorosulfonated Polyethylene	9.2	5	8.2
Polyacrylate 9.3	20	9.2
Chlorinated Polyethylene	8.7	10	8.6
Neoprene 8.2 - 9.7	0	7.8
Natural Rubber	7.9 - 8.4	2	8.0
Styrene-Butadiene Rubber	8.1 - 8.6	2	8.0
Silicone 7.3 - 8.5	0	7.8
Vynathene 9.0	20	9.2
Ethylene-Propylene Diene Terpolymer	7.9 - 8.9	0	7.8

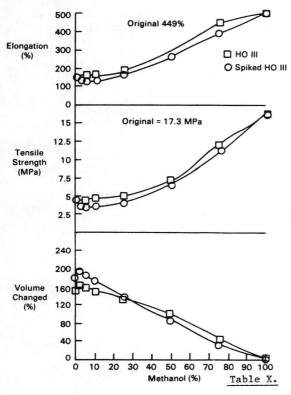

Figure 5. Tensile Elongation, Tensile Strength and Volume Change of Natural Rubber After Exposure to Methanol/ Gasoline Compositions.

Table X.

Volume Change of Fluorocarbon Elastomer (Viton A) After Immersion for 72 hr. at Room Temperature in Water and Alcohols

Alcohol	% Volume Change
Water	-1
CH_3OH	100
C_2H_5OH	2
$n-C_3H_7OH$	0.6
$n-C_4H_9OH$	0.4
$n-C_5H_{11}OH$	0.1
$n-C_6H_{13}OH$	0.1
$n-C_7H_{15}OH$	0.2
$n-C_8H_{17}OH$	0.3

To investigate whether or not the fluorocarbon elastomer behaves similarly in other alcohols, swell measurements were conducted in ethanol, propanol, butanol, pentanol, hexanol, heptanol and octanol. The volume swell results of the fluorocarbon elastomer in these alcohols and in water and methanol are shown in Table X. In all cases except methanol swelling of the elastomer is very low. In order to ascertain that diffusion rate effects are not responsible for the difference between methanol and other alcohols, the swelling of the fluorocarbon elastomer in ethanol was allowed to take place over a period of 1152 hours (48 days). This only increased the volume swell of the elastomer to 5.4% which is very much less than the 100% volume swell observed in the case of methanol.

The solubility parameters of methanol are δ_{total} = 14.5, δ_d = 7.4, δ_p = 6.0 and δ_H = 10.9 (cal/cc)$^{1/2}$. Comparing these to the solubility parameters of the fluorocarbon elastomer, the large difference is in the value of δ_H. This led us to believe that methanol in the pure form exists in a highly hydrogen bonded structure and hence has a lower δ_H value than reported. An infrared study [16] reported in the literature supported our suspicions by suggesting that methanol in pure liquid form exists in the following structure.

The structure will lead to diminished hydrogen bonding parameter as will be shown later by our NMR investigation [3] of the effects of methanol mixtures on swelling of fluorocarbon elastomers.

IV. Effect of the Degree of Swelling on Ultimate Properties of Elastomers

Equilibrium tensile properties of elastomers have been measured and found to exhibit the following dependence [17-19] on volume fraction of rubber in the swelled network:

$$\sigma = \frac{f}{A_o} = f^* (\lambda - \lambda^{-2}) \, v_r^{-1/3}$$

σ is the tensile strength , f is the force or stress, A_o is the cross-sectional area of unswollen, undistorted sample and $\lambda = \frac{L}{L_o}$ is the deformation ratio where L and L_o are the lengths of the stretched and unstretched samples respectively. f* is the reduced force and is equal to:

$$f^* = (\frac{\upsilon kT}{V}) \, [\, \langle r^2 \rangle_i / \langle r^2 \rangle_o \,] + 2 \, C_2 \, \lambda^{-1}$$

υ is the number of network chains, k is the Boltzman constant, T is the absolute temperature, V is the volume of the network, $\langle r^2 \rangle_i$ is the mean-square end-to-end distance of a network chain in the undistorted state of volume V and $\langle r^2 \rangle_o$ is the corresponding value for the chain free of network crosslinks. A plot of f* versus λ^{-1} is found to be linear at low elongations, but in the case of crystallizable polymers or polymers containing fillers an upturn is observed at higher elongations which has been explained in terms of strain-induced crystallization [20] or limited extensibility of the network [21].

All the polymers investigated in this study were filled elastomers. Our interest is to describe the effect of solvent swelling on ultimate stress-strain properties of the elastomer and to determine whether or not the effect is different for crystallizable and noncrystallizable polymers. For unfilled elastomers Mark [22] had found that both ultimate reduced stress f_b and ultimate elongation λ_b decrease with increased swell. However the rate of decrease in both properties is higher in the case of crystallizable polymers such as polyisobutylene, than noncrystallizable polymers such as polydimethylsiloxane.

Using our data the ultimate reduced stress at fracture

$$f_b^* = \frac{\sigma_b \, v_r^{1/3}}{\lambda_b - \frac{1}{\lambda_b^2}}$$

was calculated and plotted versus the fraction of rubber in the swelled network as shown in Figure 7 for nitrile, natural rubber and EPDM elastomers. A linear behavior is observed. The values of v_r in the figure are always less than one even for the unswelled elastomers because as we mentioned earlier all elastomers in this investigation contained filler. A linear regression analysis was conducted for f_b^* versus v_r for all elastomers investigated as shown in Table XI. In the case of nine of the polymers confidence limit R^2 values of greater than 0.85 are observed. This means that in the case of these polymers a linear relationship between f_b^* and v_r adequately represents the data. The slope values in Table XI indicate the degree by which the reduced stress f_b^* is influenced by the degree of swelling. Values of slope between 2.27 for silicone and 75.5 for chlorosulfonated polyethylene are obtained. Natural rubber and neoprene both of which are strain crystallizable exhibit low slope values of 3.34 and 4.18. It is obvious from the data that for filled elastomers the decrease in ultimate reduced stress with swelling is not a function of the ability of the polymer to crystallize under strain.

Simpler relationships were attempted to relate ultimate stress σ_b and percent elongation E_b to the degree of swelling. Plots of E_b and σ_b versus v_r are shown in Figure 8 for nitrile, natural rubber and EPDM elastomers. Linear regression analysis of σ_b versus v_r for all the polymers is shown in Table XI. For 14 of the sixteen polymers the confidence coefficient R^2 of 0.85 or greater shows that σ_b and v_r are linearly correlatable. Similarly a linear relationship between E_b and v_r adequately represents the data for 13 of the polymers investigated as seen again in Figure 8 and Table XI.

Logarithmic correlations of σ_b with v_r and E_b with v_r were also examined as seen in Figure 9, for nitrile, natural rubber and EPDM and in Table XI for all the polymers. From the comparison of Figures 8 and 9 and R^2 values of Table XI it is difficult to decide whether the relationships

$$\sigma_b = a + b \, v_r$$
$$E_b = c + d \, v_r$$

or

$$\ln \sigma_b = a' + b' \ln v_r$$
$$\ln E_b = c' + d' \ln v_r$$

better represent our data. In all cases strong linear correlations are observed. Values of b and d show the degree of dependence of σ_b and E_b on v_r. Again there does not seem to be a relationship between the amount of decrease of σ_b and E_b due to swelling and the ability of the polymer to strain crystallize.

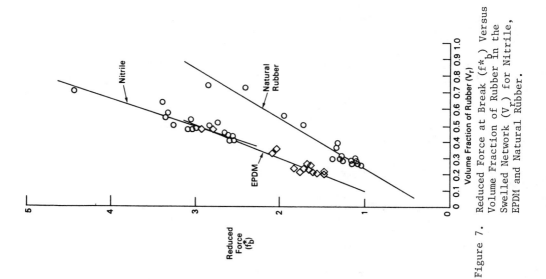

Figure 7.   Reduced Force at Break ($f*_b$) Versus
Volume Fraction of Rubber in the
Swelled Network (V_r) for Nitrile,
EPDM and Natural Rubber.

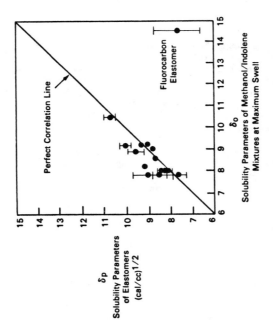

Figure 6.   Solubility Parameters of Elastomers Compared
to Solubility Parameters of Methanol/Indolene
Mixtures at Maximum Swell.

TABLE XI.

Linear Regression Analysis of Tensile Strength and Ultimate Elongation versus Volume of Rubber in the Swelled Network for Various Elastomers

Elastomer	f_b^* vs. V_r			σ_b vs. V_r			E_b vs. V_r			$\ell n\ \sigma_b$ vs. $\ell n\ V_r$			$\ell n\ E_b$ vs. $\ell n\ V_r$		
	Slope	Intercept	R^2	Slope	Intercept	R^2	Slope	Intercept	R^2	Slope	Intercept	R^2	Slope	Intercept	R^2
Fluorocarbon	8.58	-1.74	0.83	30.5	-7.70	0.90	359	-40.0	0.94	1.82	3.24	0.93	1.28	5.82	0.97
Polyether	7.29	-1.58	0.77	28.0	-8.65	0.87	534	-142	0.81	2.44	3.34	0.88	2.06	6.23	0.83
Polyester Urethane	6.38	-1.72	0.94	53.1	-21.8	0.92	815	-179	0.87	2.36	3.53	0.92	1.39	6.45	0.84
Fluorosilicone	11.7	-3.91	0.53	37.5	-11.9	0.67	973	-231	0.70	3.32	4.23	0.64	2.20	7.08	0.72
Epichlorohydrin Copolymer	2.54	1.65	.76	19.0	1.37	0.92	480	-34.0	0.91	0.86	2.98	0.90	1.14	6.11	0.88
Polysulfide	6.01	-1.23	0.86	28.5	-8.45	0.93	533	-97.8	0.87	2.24	3.30	0.90	1.50	6.17	0.84
Butadiene-Acrylonitrile (Nitrile)	6.02	0.0308	0.84	32.7	-8.29	0.93	511	-128	0.96	1.82	3.32	0.96	1.84	6.10	0.97
Chlorosulfonated Polyethylene	25.5	-4.08	.88	65.1	-12.5	0.94	271	-16.5	0.72	2.10	4.50	0.90	1.33	5.74	0.71
Polyacrylate	3.43	-0.061	0.96	17.4	-3.24	0.88	488	-77.2	0.88	1.68	2.68	0.92	1.46	5.98	0.89
Chlorinated Polyethylene	4.12	0.672	0.87	28.5	-2.46	0.95	618	-54.8	0.96	1.24	3.33	0.97	1.30	6.44	0.97
Neoprene	4.18	0.697	0.95	29.6	-2.40	0.97	589	-34.5	0.99	1.24	3.37	0.97	1.17	6.37	0.98
Ethylene-Propylene-Diene Terpolymer (EPDM)	5.01	0.537	0.92	31.0	-2.42	0.95	622	-47.8	0.97	1.28	3.43	0.93	1.29	6.46	0.96
Natural Rubber	3.34	0.206	0.95	28.4	-3.64	0.98	792	-58.7	0.91	1.48	3.35	0.97	1.33	6.74	0.93
Vynathene	3.49	-0.177	0.88	27.0	-4.03	0.92	880	-82.2	0.96	1.56	3.06	0.92	1.39	6.79	0.98
Styrene-Butadiene Rubber (SBR)	7.58	-0.966	0.58	37.4	-6.94	0.77	710	-107	0.85	1.66	3.40	0.70	1.56	6.50	0.88
Silicone	2.27	0.313	0.70	11.51	-0.282	0.86	502	-27.5	0.98	1.02	2.37	0.87	1.17	6.20	0.97

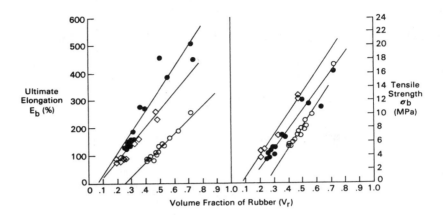

Figure 8.   Ultimate Elongation and Tensile Strength Versus Volume
             Fraction of Rubber in the Swelled Network (V_r) for ○ Nitrile,
             and ◇ EPDM, and ● Natural Rubber.

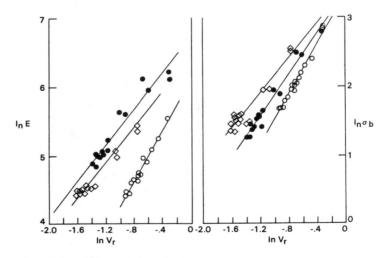

Figure 9.   Natural Logarithm of Ultimate Elongation (ln E) and Tensile
             Strength (ln σ_b) Versus Natural Logarithm of Volume Fraction
             of Rubber in the Swelled Network (ln V_r) for ○ Nitrile,
             ◇ EPDM, and ● Natural Rubber.

V. Relationship Between Tensile Strength and Ultimate Elongation

The ultimate tensile strength of an elastomer σ_b and the ultimate extension
ratio λ_b depend on the chemical and topological characteristics of polymer network as well as the conditions (for example temperature and strain rate) under
which the experiment is conducted [23]. The ultimate properties are often characterized by a failure envelope defined by values of σ_b and λ_b determined at
various strain rates over a wide range of temperatures. The resulting failure
envelope will then be independent of strain rate and temperature and will depend
only on basic characteristics of the polymeric material [23].

The equation of Martin, Roth, and Stiehler [24], shown below, represents the
equilibrium stress–strain behavior of elastomers.

$$\sigma = M \ (\varepsilon/\lambda^2) \ \exp. \ A \left[\lambda - \frac{1}{\lambda}\right]$$

In this equation M is the equilibrium tensile modulus, λ is the extension ratio
which equals $\varepsilon + 1$, and A is an adjustable parameter. We examined our ultimate
tensile data obtained on the different elastomers swelled to varying degrees in
the mixed methanol/gasoline mixtures to find out if they lie along the equilibrium stress–strain curve. For ultimate properties the above equation can be
arranged as follows:

$$\ln \frac{\sigma_b \ \lambda_b^2}{\varepsilon_b} = \ln M + A \ (\lambda_b - \frac{1}{\lambda_b})$$

Plots of $\ln \dfrac{\sigma_b \ \lambda_b^2}{\varepsilon_b}$ versus $(\lambda_b - \frac{1}{\lambda_b})$ for natural rubber and nitrile elastomers
are shown in Figure 10. It is seen that the ultimate tensile data fits the
above equilibrium stress–strain relationship well. In fact as seen from R^2
values in Table XII the data for all polymers can be well represented by the
above Martin, Roth and Steihler relationship. The intercept values, according
to the equation, should be a measure of modulus at ultimate extension. The
slope values vary between 0.516 and 1.13. For styrene–butadiene rubber
Smith [25] had calculated a value of 0.5 for this adjustable parameter A. He
obtained the various values of σ_b and λ_b for one polymer by conducting measurements at various temperatures and strain rates. In the present study a similar
value of A of 0.6 was determined by conducting the tensile measurements at constant temperatures and strain rates. The various values of σ_b and λ_b were
obtained by swelling the polymer to varying degrees in mixed solvents.

A simpler representation of the data is to relate the actual stress $\sigma_b \ \lambda_b$ to the
ultimate engineering strain [26] ε_b. Plots of the $\ln \sigma_b \ \lambda_b$ versus $\ln \varepsilon_b$ for
natural rubber and nitrile rubber are shown in Figure 11. Comparing this figure
with Figure 10 and comparing the R^2 values in Table XII, it is obvious that this
simple empirical relationship of $\ln \sigma\lambda$ vs $\ln \varepsilon$ represents the data better than
the equilibrium stress–strain relationship of Martin, Roth and Steihler [24].
The intercept of the $\ln \sigma\lambda$ vs $\ln \varepsilon$ relationship has been shown by Smith [26] to
be related to the tensile modulus. No upturn in the $\ln \sigma\lambda$ vs $\ln \varepsilon$ curve is
observed for the unswelled samples. An upturn had been observed for unfilled
natural rubber at high elongation and attributed to strain–induced crystallization [26].

CONCLUSIONS

1. Properties of most fuel resistant elastomers are degraded to a larger extent
    by mixtures of methanol and gasoline rather than by the pure components.

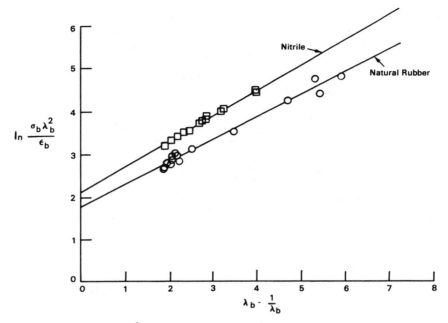

Figure 10.   ln $\dfrac{\sigma_b\lambda_b^2}{\epsilon_b}$ Versus $\lambda_b - \dfrac{1}{\lambda_b}$ for Nitrile and Natural Rubber, as per Martin, Roth, and Steihler Equation [24].

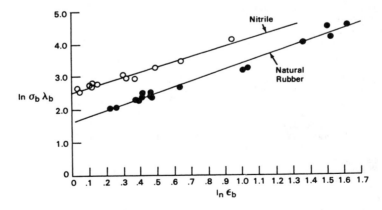

Figure 11.   Nautral Logarithm of Ultimate Actual Stress $(\ln\sigma_b\lambda_b)$ Versus Natural Logarithm of Ultimate Stretch Ratio $(\ln\epsilon_b)$ for Nitrile and Natural Rubber.

Table XII.

Linear Regression Analysis of Tensile Strength
versus Ultimate Elongation of Swelled Elastomers

Elastomer	$\ln\left(\dfrac{\sigma_b \lambda_b^2}{\varepsilon_b}\right)$ vs. $\lambda_b - \dfrac{1}{\lambda_b}$			$\ln \sigma_b \lambda_b$ vs. $\ln \varepsilon_b$		
	Slope	Intercept	R^2	Slope	Intercept	R^2
Fluorocarbon	0.910	1.58	0.94	1.96	2.24	0.95
Polyether	0.769	1.66	0.88	1.62	2.12	0.93
Polyester Urethane	0.516	2.05	0.96	2.36	1.08	0.97
Fluorosilicone	0.740	1.04	0.78	2.00	1.24	0.83
Epichlorohydrin Copolymer	0.456	2.64	0.98	1.41	2.50	0.99
Polysulfide	0.758	1.47	0.92	2.02	1.70	0.95
Butadiene-Acrylonitrile (Nitrile)	0.665	2.24	0.97	1.55	2.54	0.98
Chlorosulfonated Polyethylene	1.13	2.34	0.81	1.70	3.34	0.90
Polyacrylate	0.773	1.23	0.97	1.64	1.83	0.98
Chlorinated Polyethylene	0.561	2.21	0.98	1.60	2.23	0.99
Neoprene	0.590	2.15	0.99	1.73	2.14	0.99
Ethylene-Propylene-Diene Terpolymer (EPDM)	0.672	1.96	0.98	1.58	2.30	0.99
Natural Rubber	0.521	1.77	0.98	1.78	1.62	0.98
Vynathene	0.583	1.10	0.99	1.77	1.30	0.97
Styrene-Butadiene Rubber (SBR)	0.611	1.82	0.93	1.75	2.07	0.93
Silicone	0.572	1.48	0.94	1.48	1.69	0.96

2. The data on all elastomers except the fluorocarbon can be explained in terms of the solubility parameter concept.

3. The ultimate tensile strength and ultimate elongation of swelled elastomer networks are quantitatively related to volume swell by simple linear relationships.

4. Ultimate stress and ultimate elongation of swelled elastomer networks obey equilibrium stress-strain relationships.

ACKNOWLEDGEMENT

It is a pleasure to acknowledge the capable technical assistance of Ms. Dorthie McIntyre in obtaining part of the data and conducting computer analyses. The assistance of Dr. R. J. Salloum in computer programing and Mr. John A. Krohn in obtaining most of the data is also gratefully acknowledged.

REFERENCES

1. I. A. Abu-Isa, SAE Technical Paper Series No. 800786, Passenger Car Meeting, Dearborn, MI June 9-13, 1980.

2. "Octel Group Gasoline Survey, Europe," Associated Octel Company Limited, London, July 1977.

3. I. A. Abu-Isa and M. E. Myers "Elastomer/Gasoline blends Interactions III. Investigation of the Effects of Methanol on Fluorocarbon Elastomers" in preparation.

4. C. M. Hansen, J. Paint Tech. $\underline{39}$, 104 (1967).

5. C. M. Hansen and K. Skaarup, J. Paint Tech. $\underline{39}$, 511 (1967).

6. H. Ahmad and M. Yaseen, Polym. Eng. and Sci. $\underline{19(12)}$, 858 (1979).

7. J. L. Hildebrand and R. L. Scott "The Solubility of Nonelectrolytes" Reinhold Publishing Company, New York, 1949.

8. P. J. Flory and J. Rehner, Jr., J. Chem. Physics $\underline{11}$, 521 (1943).

9. P. J. Flory, J. Chem. Physics $\underline{18}$, 108 (1950).

10. N. R. Langley and J. D. Ferry, Macromolecules $\underline{1(4)}$, 353 (1968).

11. C. J. Sheehan and A. L. Bisio, Rubber Chem. and Tech. $\underline{39(1)}$, 149 (1966).

12. A. Beerbower and J. R. Dickey, ASLE Trans. $\underline{12}$, 1 (1969).

13. A. F.M. Barton, Chem. Reviews $\underline{75(6)}$, 731 (1975).

14. I. A. Abu-Isa, Research Publication GMR-3378 "Radiation Curing of Chlorinated Polyethylene", to be published in the Journal of Radiation Curing.

15. J. D. MacLachlan, Polym. Plast. Tech. Eng. $\underline{11(1)}$, 41 (1978).

16. D. N. Glew and N. S. Rath, Canadian Journal of Chemistry 49, 837 (1971).

17. T. J. Dudekand and F. Bueche, Rubber Chem. and Tech. 37, 894 (1964).

18. P. J. Flory "Principles of Polymer Chemistry", Cornell University Press, Ithaca, N.Y. 1953.

19. J. E. Mark, Rubber Chem. and Tech. 48, 495 (1975).

20. J. E. Mark, Polymer Engineering and Science 19(4), 254 (1979).

21. W. O. S. Doherty, K. L. Lee and L. R. G. Treloar, The British Polymer Journal 12(1), 19 (1980).

22. J. E. Mark, Polymer Engineering and science 19(6), 409 (1979).

23. T. L. Smith, Trans. Soc. Rheology 6, 61 (1962).

24. G. M. Martin, F. L. Roth and R. D. Stiehler, Trans. Inst. Rubber Ind. 32, 189 (1956).

25. T. L. Smith, J. Polym. Sci. Part A 1, 3597 (1963).

26. T. L. Smith, J. Appl. Physics 35(1), 27 (1964).

GLOSSARY

Polymers

CM 0136	Chlorinated polyethylene, 36% chlorine, Dow chemical.
F-70A	Polyether elastomer, American Cyanamid.
Herclor C	Epichlorohydrin-ethylene oxide copolymer, Hercules.
Hypalon 40SS	Chlorosulfonated polyethylene, 35% chlorine, 1% sulfur, DuPont.
Hypalon 48	Chlorosulfonated polyethylene, 43% chlorine, 1% sulfur, DuPont.
Hycar 1042	Butadiene-acrylonitrile copolymer, medium-high acrylonitrile content, B. F. Goodrich Chemical.
Hycar 4042	Acrylic ester copolymer, sp. gr. 1.10, B.F. Goodrich Chemical.
Neoprene W	Polychloroprene, Mooney Visc. 50, DuPont.
Polysulfide FA	Polysulfide, sp. gr. 1.34, Thiokol Chem.
SBR 1500	Styrene-butadiene rubber, 23.5% styrene, B. F. Goodrich.
Silastic LS-40	Fluorosilicone, sp. gr. 1.38, durometer when cured 40 shore A points. Dow Corning.

Silastic LS-70	Fluorosilicone, sp. gr. 1.48, durometer when cured 70 shore A points. Dow Corning.
Silicone GP-70	General purpose silicone with methyl and vinyl attached groups, 70 durometer. Dow Corning.
Smoked Sheet	Coagulated natural rubber sheets, properly dried and smoked, Standard Malaysian Rubber.
Vibrathan 5004	Millable polyurethane polymer, Uniroyal Chemical.
Vistalon 6505	Rapid curing, sulfur curable ethylene-propylene-diene terpolymer, Exxon.
Viton AHV	Copolymer of vinylidene fluoride and hexafluoropropylene, DuPont.
Vynathene EY904	Vinyl acetate-ethylene copolymer, sp. gr., 0.98, U.S. Industrial Chemicals.

Crosslinking Agents

Altax	Benzothiazyl disulfide
Captax	2-mercaptobenzothiazole
Curative p	Proprietary chemical made by American Cyanamid Company specifically for curing F-70A polyether elastomer.
Diak #1	Hexamethylene diamine carbamate
Di-Cup	Dicumyl peroxide
DPG	Diphenyl-guanidine
HVA-2	N,N-m-phenylenedimaleimide
Maglite D	Magnesium oxide
MBTS	2-Mercaptobenzothiazole
Methyl Tuads	Tetramethylthiuram disulfide
NA-22	2-Mercaptoimidazoline
Protox 166	Zinc oxide surface treatd with propionic acid.
Santocure	N-Cyclohexyl-2-benzothiazolesulfenamide
TAIC	Triallylisocyanurate
Tetrone A	Dipentamethylene thiuram
TMPT	Trimethylolpropane trimethacrylate

Varox	2,5-bis (tert-butyl peroxy)-2-5-dimethylhexane.

Oils and Plasticizers

A-C PE617A	Low molecular weight polyethylene
Chlorowax LV	Low viscosity chlorowax containing about 40% chlorine.
Paraplex G-62	Soybean oil epoxide.
Reogen	Sulfonic acid and paraffin oil blend
TP 759	Polymeric plasticizer from Thiokol Chemical

Stabilizers and Antioxidants

AgeRite DPPD	Diphenyl-p-phenylene-diamine
AgeRite HP	A blend of dioctylated diphenylamines and diphenyl-p-phenylenediamine
AgeRite Resin D	Polymerized 1,2-dihydro-2,2,4 trimethyl quinoline
Antozite 67	N-(1,3-dimethylbutyl)-N'-phenyl-phenylenediamine
Dythal XL	Dibasic lead phthalate
NBC	Nickel Dibutyldithiocarbamate
Octamine	Reaction product of diphenylamine and diisobutylene
Silastic HT-1	Proprietary heat stabilizer for silicone rubber, Dow Corning
Sunpar 2280	Paraffinic oil, Sp. gr. 0.8916, aromatic content 23.5%
TLD-90	90% Lead oxide dispersed in a 10% polymer matrix

NEW EMPIRICAL RELATION BETWEEN COHESION ENERGY DENSITY AND ONSAGER REACTION FIELD FOR SEVERAL CLASSES OF ORGANIC COMPOUNDS*

Miomir B. Djordjevic and Roger S. Porter

Polymer Science and Engineering Department
Materials Research Laboratory
University of Massachusetts
Amherst, Massachusetts 01003

ABSTRACT

For some organic compounds cohesive energy density, CED, within a homologous series is found to change linearly with a single material parameter, $g^2 = [(n^2 - 1)/(2n^2 + 1)]^2$, reflecting the intensity of the Onsager reaction field; n is the refractive index of the compound. This correlation has been found for n-alkanes, 1-n-alkenes and n-alkyl benzenes (exhibiting only dispersive interactions), for methyl-n-alkyl ketones (with dipolar interactions) and for linear alcohols (forming hydrogen bonds). Correlation for n-alkanes, extrapolated to corresponding g^2 is in good agreement with CED for polyethylene.

Substituent changes both g^2 and CED, of the compound, relative to the corresponding n-alkane. Change of g^2 reflects the effect of the substituent on purely dispersive interaction. The contribution of the specific interactions (dipolar or hydrogen bonds) causes the difference between CED for a substituted n-alkane and an imaginary n-alkane of equal g^2. For methyl-n-alkyl ketones this contribution is in good agreement with previously reported values.

KEYWORDS

Cohesive energy density; Onsager reaction field; refractive index; intermolecular interaction; dispersive interaction; dipolar interaction; hydrogen bond; alkanes; alkyl benzenes; ketones; alcohols.

INTRODUCTION

Cohesive energy density, CED (1), represents the cumulative strength of intermolecular interactions within a pure compound in the condensed state. An interest thus exists both in determining the contributions of each interaction type to the total CED and in correlating the CED with other material properties.

For compounds devoid of strong interactions, cohesive interactions may be considered to be the sum of dispersive (nonspecific) interaction (2), dipole-dipole (3) and

*Dedicated to Joel H. Hildenbrand, the founder of the concepts of intermolecular interaction, that are the basis for this work, on the occasion of his 100th. birthday, November 16, 1981.

dipole-induced dipole interactions (4,5). The contribution of each interaction type to the total CED may be estimated as the sum of interaction energies for pairs of molecules whose physical properties, such as: dipole moment (6), polarizability (7) and ionization potential (8), are known. These approaches can be applied, however only as approximations (1). This subject has been reviewed by Hildenbrand and Scott (1), by Gardon (9) and by Pitzer (10). A review (11) is also available of the several approaches (12-14) for calculation of contributions to the CED of strong intermolecular interactions, such as hydrogen bonds.

Correlations between CED and other molecular properties are of double importance. They may permit an indirect determination of CED, and, more important, may help understand the mechanisms of interaction.

The refractive index, n (15), is one of several material parameters that can be correlated to other physical properties. Several efforts have been made to develop correlations (16-20) between CED and the Lorentz-Lorenz function (21). Generally attempts were made to correlate values for a broad array of compounds of different chemical types. Results apparently do not reveal any mechanism which would be common for the cohesive interactions of all the compounds.

Significant theoretical effort has been oriented toward understanding the relation between the evaporation enthalpy of organic compounds and the intensity of the Onsager reaction field, ORF (7,22-25)

$$\underset{\sim}{R} = \frac{2}{a^3} \; (\frac{\varepsilon - 1}{2\varepsilon + 1}) \; \underset{\sim}{\mu} \tag{1}$$

(a is the diameter of Onsager's spherical cavity, enclosing the dipole of moment, $\underset{\sim}{\mu}$, and ε is the bulk dielectric constant of the compound [22]). Theoretical results indicate a functional dependence of cohesive energy on ORF (7,23-25). However, the agreement between theoretically calculated and experimentally determined values has been only qualitative. A possible limitation in those theoretical approaches may be that a single relation was attempted for an array of unrelated and structurally dissimilar compounds.

To our knowledge, there has been no prior published attempts to correlate experimental values of CED (or quantities reflecting the intensity of cohesive interaction) with any of the parameters characterizing the intensity of ORF for a range of homologous organic compounds.

METHOD

By introducing the Maxwell relation (26)

$$\varepsilon = n^2 \tag{2}$$

[where $n = n_{D,\,25}$ is the refractive index of the compound for the Na D line (5992.6$\overset{\circ}{A}$) at 25°C (21)] the intensity of the ORF can be expressed as

$$\underset{\sim}{R} = \frac{2}{a^3} \; (\frac{n^2 - 1}{2n^2 + 1}) \; \underset{\sim}{\mu} \tag{3}$$

or

$$\underset{\sim}{R} = g \; \frac{2\underset{\sim}{\mu}}{a^3} \tag{3a}$$

where the reaction field factor

$$g = \frac{n^2 - 1}{2n^2 + 1}$$

is a dimensionless material parameter (7). It represents the effect of the continuum on the ORF induced by the permanent dipole imbedded in it (22). It has been demonstrated earlier that expressions (3a) and (4) also apply to the intensity of ORF induced by the fluctuating dipole in a nonpolar compound (23).

CED for the compound can be calculated from the enthalpies of evaporation at the equilibrium vapor pressure at 25°C (9,27). The specific choice of the compounds whose characteristics were correlated in this analysis has been dictated by the availability of the reliable enthalpic data for the homologue series (9,27). Compounds selected have been classified in six homologous series (Table 1) and in at least three categories:
(a)   n-alkanes, 1-n-alkenes and n-alkyl benzenes, that exhibit only dispersive interactions (1)
(b)   ketones, that in addition exhibit dipolar interactions (1,28,29), and
(c)   alcohols, that may further interact through hydrogen bonding (1,30).

Linear alkyl segments are present in every compound, their length being the difference between members in each series, and all of these compounds may be represented as a linear chain without a bulky side group. Compounds with different chemical properties, but with an equal number of atoms in the chain are homomorphous, similar in both shape and size. One example is the group n-hexane, 1-n-hexene, n-pentanol and 2-hexanone.

RESULTS

TABLE 1 Cohesion Energy Densities and the Onsager Reaction Field Parameters for Several Homologous Organic Series at 25°C and 1 atm.

No	Compound	g^2	CED, cal/cm^3	Reference
	n-alkanes - correlation coefficient[a] 0.9990			
1	propane	0.0235	40.96[b]	14
2	butane	0.0290	46.24[b]	14
3	pentane	0.0320	49.22	12
4	hexane	0.0343	52.80	12
5	heptane	0.0361	55.22	12
6	octane	0.0374	57.01	12
7	nonane	0.0385	58.79	12
8	decane	0.0394	59.63	12
9	undecane	0.0404	61.00[c]	13
10	dodecane	0.0411	62.73[c]	13
11	tetradecane	0.0419	63.20[c]	13
	1-alkenes - correlation coefficient 0.9988			
12	pentene	0.0338	49.18	12
13	hexene	0.0360	49.33[c]	12
14	heptene	0.0377	55.92	12
15	octene	0.0389	57.69	12
16	nonene	0.0399	59.10	12
17	decene	0.0407	60.24	12
	Methyl-n-alkyl ketones - correlation coefficient 0.9945			
18	2-propanone	0.0325	95.43	13
19	2-butanone	0.0351	85.93	13
20	2-pentanone	0.0367	79.57	13
21	2-hexanone	0.0381	74.30	13
22	2-heptanone	0.0394	72.25	13
23	2-octanone	0.0401	71.40[c]	13

No	Compound	g^2	CED, cal/cm^3	Reference
	Benzene and alkyl substituted benzenes - correlation coefficient 0.9885			
24	benzene	0.0514	83.86	12
25	toluene	0.0509	79.43	12
26	ethylbenzene	0.0507	77.23	12
27	n-propylbenzene	0.0502	74.63	12
	Linear alcohols with odd number of carbons - correlation coefficient 0.9885			
28	methanol	0.0283	210.25	13
29	propanol	0.0359	141.61	13
30	pentanol	0.0391	118.81	13
31	heptanol	0.0411	100.00	13
	Linear alcohols with even number of carbons - correlation coefficient 0.9999			
32	ethanol	0.0326	161.29	13
33	butanol	0.0377	129.96	13
34	hexanol	0.0403	114.49	13
35	octanol	0.0418	106.09	13

(a) Since not all conditions are known for the experimental determination of these material parameters, linear correlation coefficients were determined only for data from one source (as indicated by reference).
(b) Refractive index determined at different pressures and temperatures. Experimental conditions for the determination of the CED are unknown. Points not considered in correlation.
(c) Points not considered in correlation.

Relations expressed in Table 1 and in Fig. 1.1 to Fig. 1.3 indicate that within each homologous series, CED depends linearly on the single material parameter, g^2

$$g^2 = (\frac{n^2 - 1}{2n^2 + 1})^2 \qquad (5)$$

For n-alkanes, see Fig. 1.1, CED is proportional to g^2. Experimental points for alkanes of low molecular weight deviate from the correlation. The upward orientation of this deviation suggests that it may be caused by the errors in determining the small enthalpies of evaporation for those compounds. Assuming that polyethylene, PE, is ideally linear the observed linear dependence of CED on g^2 for n-alkanes can be extrapolated to the g^2 value for PE. Literature contains different values of n for PE. Using values from two sources, g^2 has been calculated as 0.050 (31) and 0.053 (32). Extrapolating linear dependence of CED on g^2 for n-alkanes corresponding values of CED for PE are 75.5 and 79.2 cal/cm^3, respectively. These values are 13 to 27% higher than reported experimental results (62 to 66 cal/cm^3) (9,32).

The agreement between the extrapolated and experimental CED may be considered acceptable for an extrapolation based on a single material parameter, and on a series of only six n-alkanes between pentane and decane. Higher values for extrapolated CED may be attributed to the simpler structure of short n-alkanes. The size of these differences, however, indicate that the interactions making the main contribution to CED in PE have the same character as in low molecular weight n-alkanes.

For linear 1-n-alkenes, Fig. 1.1, CED is also increasing with g^2. The 1-n-alkenes have both g^2 and CED higher than homomorphous n-alkanes, yet, their CED are smaller than for imaginary n–alkanes of corresponding g^2, see Fig. 1.1. Those differences are decreasing with molecular weight.

Similarly, n–alkyl benzenes, see Fig. 1.2, have CED increasing linearly with g^2 with a slope which is much steeper than for n–alkanes. Short chain substituted benzenes have CED values higher than n–alkanes with corresponding g^2, see Fig. 1.1 and Fig. 1.2, but, different from the cases of n–alkanes or 1–n–alkenes, both g^2 and CED decrease as the length of n–alkyl segment increases. Interestingly, CED for benzene falls on the same line for other members of the series.

For the polar compounds, methyl n–alkyl ketones, see Fig. 1.2, and linear alcohols, see Fig. 1.3, the CED decrease linearly with g^2. It is noteworthy that separate linear correlations exist for alcohols with an odd and for alcohols with an even number of carbons. Each of the correlations for the polar compounds includes the smallest member of the series: acetone, methanol and ethanol, respectively. In all cases CED decrease with the increase in the size of n–alkyl segments.

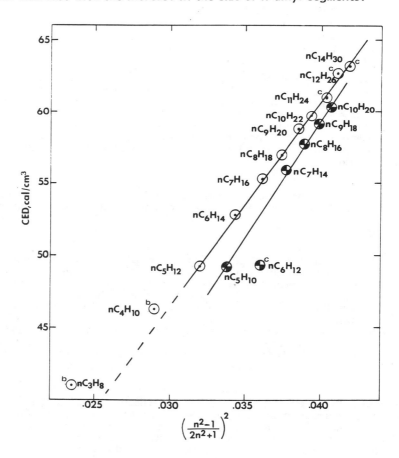

Fig. 1.1. Change of CED with g^2 for n–alkanes (0) and 1–n–alkenes (●) at 25°C; superscripts correspond to those in Table 1.

DISCUSSION

In several previous works, attempts have been made to correlate CED with easily accessible parameters representing the intensity of ORF. Theoretical findings have suggested the dependence of the enthalpy of evaporation on g (7,23). In our work the best correlation has been obtained when CED was compared with g^2. The remarkable feature here is that unsubstituted n-alkanes and substituted, polar and nonpolar compounds all show a linear dependence of their CED on the same parameter, g^2. The fact that compounds with small molecules and pronounced dipolar or hydrogen bonding interactions correlate with other members of the series, see Fig. 1.2 and Fig. 1.3, is unexpected and requires explanation.

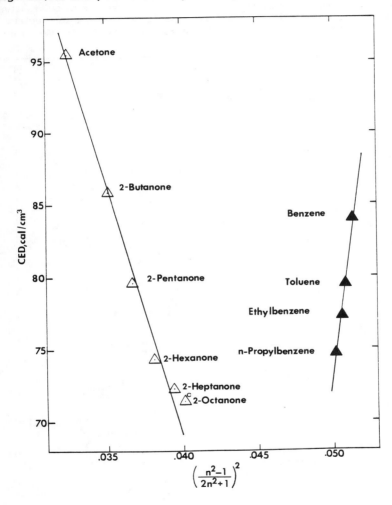

Fig. 1.2. Change of CED with g^2 for methyl-n-alkyl ketones (Δ) and n-alkyl benzenes (▲) at 25°C; superscript corresponds to this in Table 1.

Accepting the homomorphous compound approach (33) the difference in CED between an n-alkane and its homomorph, 1-substituted n-alkane, may be considered a contribution of the substituent to CED. This contribution is usually assessed by comparing the CED of the two homomorphous compounds at equal reduced temperatures (33). Results reported in this work indicate that a new approach is possible for assessing the effect of a substituent on CED. Apparently the effect of the substituent is at least twofold:

(a) For all cases studied here, the introduction of the substituent increases considerably the value of the ORF parameter g^2, over the value for an homomorphous n-alkane, and even more over the value for an n-alkane with an equal number of carbon atoms. It has been demonstrated for a nonpolar molecule in polar solvents (34) and vice versa (29) that ORF is a good representative of the total electric field on the solute. The intensity of purely dispersive, nonspecific, intermolecular interac-

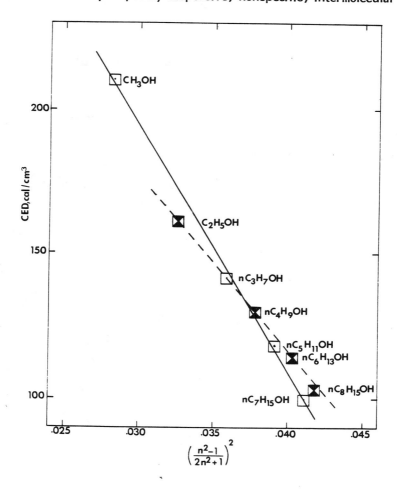

Fig. 1.3. Change of CED with g^2 for linear alcohols with odd (□) and even (▨) number of carbons.

tion between a 1-substituted n-alkane and the continuum may thus be expected to be equal to the intensity of interaction between its homomorphous n-alkane and the medium of equal g^2. In the first approximation it may be equated to the intermolecular interaction between one molecule and the continuum in pure n-alkane with the same g^2. Thus the contribution of purely dispersive interaction to the total CED at 25°C corresponds to the CED of an imaginary n-alkane with g^2 equal to that of the substituted compound.

(b) The substituent may enter in specific intermolecular interactions (2–5,28,30). This increases CED and is reflected in the difference between CED of the substituted n-alkane and the imaginary n-alkane with equal g^2. In this work the contribution of intermolecular interactions, other than dispersive to CED of methyl-n-alkyl ketones at 25°C are 2 to 18% higher than values reported earlier for the same set of compounds and for temperatures at which their vapor pressure is 100 mm/Hg (35).

For n-alkanes, 1-n-alkenes and n-alkyl benzenes the intensity of CED increases with the rise of g^2. This is in an agreement with the assumption that those compounds interact only through dispersive interaction. For n-alkanes, this indicates also that all the molecular parameters determining the intensity of interaction of the molecule with the field are independent of molecular weight. This in turn implies that the methylene group is the structural unit involved in intermolecular interaction. In the case of 1-n-alkenes the vinyl group enhances g^2 but at the same time CED is below that of an imaginary n-alkane with equal g^2. This difference is smaller for longer molecules. A possible explanation may be associated with the density reduction due to the stiff vinyl group. The change of CED with g^2 for benzene and n-alkyl benzenes suggests that the character of intermolecular interaction may not be purely dispersive as customarily assumed, or that the aromatic ring sensitizes the compound to the effect of ORF. Following this argument it will be interesting to see whether the higher members of the series will have CED below the level of n-alkanes with the corresponding g^2. The lack of experimental values for their evaporation enthalpies prevents, however, a further analysis.

Both alcohols and ketones show the decrease of CED with increase of g^2. For alcohols the fraction of CED above the dispersive interaction may be attributed to the hydrogen bonding (1,30). For ketones it may be attributed to strong dipole-dipole interactions (1,28,29). In both cases it may be noticed that this decrease follows the increase in the molar volume, i.e. the decrease of the equivalent concentration of groups able to enter in specific interactions. Assuming the validity of the isomorphous concept this indicates that only the contribution of the specific interactions to total CED decreases as the size of the n-alkyl segment increases.

CONCLUSIONS

1. For nonpolar n-alkanes and for both nonpolar and polar 1-substituted n-alkanes, the CED within an homologous series is found to change linearly with a single material parameter, $g^2 = [(n^2 - 1)/(2n^2 + 1)]^2$, reflecting the intensity of the Onsager reaction field, ORF.

2. For n-alkanes, the dependence of CED on g^2 can be extrapolated to the g^2 for the limiting member of the series, PE. CED obtained this way for two g^2, both based on values of n reported in the literature are in fair agreement with experimentally determined CED for commercially produced, branched PE.

3. The presence of either an electron-rich or a polar substituent increases the intensity of ORF and thus increases the intensity of purely dispersive interaction for the substituted n-alkanes above the level of the homomorphous n-alkanes.

4. Polar groups, as substituent, cause CED to rise above the level for n–parafins with corresponding g^2. For alcohols, this increase is attributed to the hydrogen bond and for ketones to dipole–dipole interaction. The relative contribution of specific interaction to CED decreases as the size of the molecule increases.

5. Vinyl groups, as substituent, depresses the CED below the level for n–alkanes of equivalent g^2. A possible explanation may be related to the increase in the molar volume of the liquid due to the stiff vinyl group.

6. Phenyl ring, as substituent, causes a strong increase in both g^2 and in CED for short chain n–alkyl benzenes. However the effect of phenyl ring on CED requires further experimental and theoretical analysis.

REFERENCES

1. J.H. Hildebrand and R.L. Scott, The Solubility of Nonelectrolytes, 3rd. ed., Reinhold Publishing Corp., New York, 1950.
2. F. London, Trans. Faraday Soc., 33, 8 (1437).
3. W.H. Keesom, Physik Z., 22, 643 (1921); 23, 225 (1922).
4. P. Debye, Physik. Z., 21, 178 (1920); 22, 302 (1921).
5. H. Falkenhagen, Physik. Z., 23, 87 (1922).
6. O. Exner, Dipole Moments in Organic Chemistry, Thieme, Stuttgart, Germany, 1975.
7. C.J. Bottcher, O.C. Van Belle, P. Bordewijk and A. Rip, Theory of Electric Polarization, Elsevier, New York, 1975.
8. V.I. Vedeneev, Bond Energies, Ionization Potentials and Electron Affinities, G. Arnel, London, 1966.
9. J.L. Gardon, Encyclopedia of Polymer Science, John Wiley and Sons, New York, 1965; Vol. 3, p. 833.
10. K.S. Pitzer, Advan. Chem. Phys., 2, 59 (1959).
11. A.F.M. Barton, Chem. Rev., 75, 731 (1975).
12. W. Gordy, J. Chem. Phys., 7, 93 (1939); W. Gordy and S.C. Stanford, J. Chem. Phys., 8, 170 (1940); 9, 204 (1941).
13. J.W. Crowley, G.S. Teague and J.W. Lowe, J. Paint. Technol., 38, 269 (1966).
14. R.C. Nielson, R.W. Hemwall and G.D. Edwards, J. Paint. Technol., 42, 636 (1970).
15. J.R. Partington, An Advanced Treatise on Physical Chemistry, Vol. IV, p. 1, Longmans, Green and Co., London, 1955.
16. P. Walden, Z. Phys. Chem., 70, 587 (1910).
17. G. Scatchard, Chem. Rev., 44, 7 (1949).
18. D.D. Lawson and J.D. Ingham, Nature, 223, 614 (1969).
19. R.A. Keller, B.L. Karger and L.R. Snyder, Gas Chrom. Proc., 8, 125 (1970).
20. R. Bonn and J.J. van Aartsen, Eur. Polym. J., 8, 1055 (1972).
21. J.R. Partington, An Advanced Treatise on Physical Chemistry, Vol. IV, p. 8, Longmans, Green and Co., London, 1955.
22. L. Onsager, J. Amer. Chem. Soc., 58, 1486 (1936).
23. B. Linder, Advan. Chem. Phys., 12, 225 (1967).
24. B. Linder, J. Chem. Phys., 33, 668 (1960).
25. V.A. Gorodyskii, L.F. Kardashina and N.G. Bakhshiev, Russ. J. Phys. Chem., 49, 644 (1975).
26. J.R. Partington, An Advanced Treatise on Physical Chemistry, Vol. IV, p. 424, Longmans, Green and Co., London, 1955.
27. American Petroleum Institute, Technical Data Book – Petroleum Refining, Ch. 1, Table 1c1.1.
28. H. Saito, Y. Tanaka, S. Nagata and K. Nukada, Can. J. Chem., 51, 2118 (1973).
29. B. Tiffon and J.-E. Dubois, Org. Magn. Res., 11, 295 (1978).

30. K.D. Nisbet in F.W. Harris and R.S. Seymour "Structure-Solubility Relationships in Polymers", Academic Press, New York, 1977, p. 33.
31. D.W. Van Krevelen, Properties of Polymers, 2nd. Edition, Elsevier, New York (1976), Ch. 10.
32. R.C. Weast, Ed., CRC Handbook of Chemistry and Physics, CRC Press, Inc., Boca Raton, Florida, 1980.
33. H.C. Brown, G.K. Barbaros, H.L. Berneis, W.H. Bonner, R.B. Johanesen, M. Grayson and K.L. Nelson, J. Amer. Chem. Soc., 75, 1 (1953).
34. F.H.A. Rummens, Chem. Phys. Lett., 31, 596 (1975).
35. E.F. Meyer, T.A. Renner and K.S. Stec, J. Phys. Chem., 75, 642 (1971).

DISSOLUTION KINETICS OF POLYMERS:
EFFECT OF RESIDUAL SOLVENT CONTENT

A. C. Ouano

IBM Research Laboratory, 5600 Cottle Road,
San Jose, California 95193

ABSTRACT

A study on the effect of residual solvent content on the dissolution rate of poly(methyl methacrylate) (PMMA), cresol-formaldehyde resin (Novolac), resist (a mixture of Novolac resin and adiazo-photo-active compound, (PAC)), has shown the following: (1) the dependence of the dissolution rate on the residual solvent content is very strong and that the dissolution rate - solvent content relationship can be interpreted in terms of the free volume theory; (2) the addition of the PAC to the Novolac resin decreases the residual solvent content of the resists; and, (3) very rapid drying of P(MMA) at 160°C, results in very fast dissolution rate. It takes more than 30 min of annealing to reduce the "extra free volume" created by the rapid loss of solvent at 160°C.

It is known that the diffusion rate of the solvent into polymeric matrices is very strongly dependent on the free volume and/or the solvent content of the polymer. This is a consequence of the increased mobility of the solvent molecule in the polymer matrix with free volume and solvent concentration. These results help support the hypothesis that the dissolution rate of polymers like PMMA and Novolac resins is governed by the rate of solvent diffusion into a well-annealed and of very low residual solvent content; e.g., 4%< polymer matrix. These results also explain the reduction of the solubility rate of Novolac resin with the addition of the PAC. They are also consistent with the hypothesis that the high dissolution rate of X-ray and e-beam exposed P(MMA) is due to the extra free volume created by the evolution of gaseous products and strain in the P(MMA) matrix created by irradiation.

I. INTRODUCTION

It is well known that the dissolution kinetics of a resist, be it a high molecular weight (MW) polymer, e.g., PMMA[1] or a two component system, e.g., the diazo-Novolac[2,3] type, depends to a large extent on the prebaking condition. Resist are radiation sensitive materials, and as such are used to generate lithographic patterns in the fabrication of large scale integrated circuits. Illustrative of this effect is the data[1] on the dissolution rate of P(MMA) presented in Fig. 1. Films of 3000Å initial thickness on silicon wafer substrate are completely dissolved in MIBK (methyl isobutylketone) in less than 1 min if prebaked at 100°C for 5 min, but lose <500Å in a 4 min exposure to solvent if prebaked at 160°C for 1 hour. Since P(MMA) is known to be stable at the prebaking conditions used, particularly with regard to crosslinking, the large decrease in the solubility rate with increasing time and prebake temperature has been ascribed to the quantity of residual solvent in the film.[1]

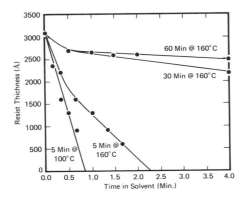

Fig. 1. Dissolution rate curves for P(MMA) in MIBK solvent at various prebaking conditions.[1]

Previously[4] it has been reported that the rate of dissolution of a glassy polymer depends strongly on the diffusion rate of the developer (solvent) into the glassy matrix. Since the diffusion coefficient of the developer depends strongly on the physical state of the cast film, i.e., the residual solvent content, strain, presence of microcracks and voids, it is hypothesized that the prebaking process serves two purposes: removing the casting solvent and annealing the film.

Based on the above, our study was to investigate the relationship between the residual solvent and the dissolution kinetics of the diazo-Novolac resist and P(MMA). It is expected that the results obtained would be of value in improving lithographic processing conditions.

II. EXPERIMENTAL

Since one of the objectives of this work is to relate the results obtained to actual lithographic processing conditions, experiments were carried out on resist films of approximately 1.5 μm thickness spin-coated on 1 inch silicon wafers. The spin coating station used was a Headway Research, Inc. Model EC101. Spinning condition was 2000 rpm for 45 secs using solutions of resist or resin containing 34% solids in diglyme solvent. The P(MMA) solution was 8% Elvacite 2041 in chlorobenzene. The coating formulations studied were: resin (100% Novolac resin), resist (82% Novolac resin and 18% photoactive compounds, PAC) and resist analog (82% Novolac resin and 18% photoactive compound analog, PACA). PACA is a photoactive compound analog whose chemical structure is identical to that of the PAC except for absence of the diazo oxide group. The generalized structures of both the PAC and PACA are illustrated below:

PAC   PACA

where R is usually a mono- (above), di- or tri-substituted diphenyl ketone. The $\overline{M}_w$ of the P(MMA) and Novolac resin used in this study were 5×10^5 and 10^4, respectively.

Since the PACA is stable to light and heat (it can be heated well above 100°C without decomposition), it allows an investigation of the effect of prebaking at high temperature (about 100°C) on the solubility rate of the resist without danger of decomposition. Thus it is possible to study the effect of residual solvent and annealing without the complication of changing the chemical nature of the resist.

The amount of retained casting solvent was determined by weighing the silicon wafer prior to spin coating and the film/wafer system prior and subsequent to prebake. Weights were recorded using a Cahn 1000 Electrobalance capable of measuring weight changes of 10^{-6} grams with reasonable accuracy.

Prebaking of coated wafers was carried out on an explosion-proof Thermoline hotplate equipped with a Powerstat to control the prebaking temperature and fitted with a heavy aluminum block heating surface to minimize temperature variations over the surface. Using thermocouples, edge to edge variation in temperature over a range of temperature (70°C to 160°C) was found to be less than 2°C. The temperature of the top surface of the P(MMA) or resist coating reached the hot plate surface temperature in less than 10 secs, because of the very thin (<0.01 in.) silicon wafer substrate. P(MMA) coatings were baked over a range of temperature and time, while resist and resin were baked for 30 min.

A brief description of the experimental procedure following the specification stated above is as follows:

(1) Solutions in diglyme of the resin, resist and resist analog were prepared in ca. 34% solid concentration. An 8% Eluacite 2041 P(MMA) solution was also prepared. Immediately prior to spin coating, these solutions were filtered through a 0.45 μm Fluoropore filter.

(2) The weights of each silicon wafer were taken prior to spin coating the resist. After spin coating, the coated wafers were immediately weighed in the Cahn Electrobalance. This gives the initial weight of a coated film.

(3) The weighed, coated wafers were subsequently baked on the hot plate at various temperatures (70°C to 160°C for the Novolac type resist and resin, and 85 to 160°C for PMMA) for various lengths of time. After allowing the wafers to equilibrate at room temperature (inside wafer carriers), they were reweighed. The difference in initial and final weights gives the weight loss as a function of prebaking temperature.

(4) After prebaking and reweighing, the initial thickness of the coatings were measured by a Taylor-Hobson Talystep, and then dissolved in a 0.257N KOH solution (MIBK for PMMA). The thickness loss as a function of time were obtained by laser interferometry and the solubility rates calculated from these data.

III. RESULTS AND DISCUSSION

Figure 2 shows the plots of weight loss versus prebake temperature for resin, resist and resist analog films coated on silicon wafers. At least two observations can be made from Fig. 2 in regard to the casting solvent (diglyme) weight loss characteristics of the sample studied.

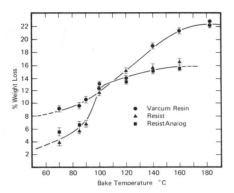

Fig. 2. Weight loss versus baking temperature plots of the resin, resist and the resist analog. Baking time was held constant at 30 min.

(1) The resin curve has a higher prebaking temperature-weight loss asymptote than both the resist and the resist analog formulations. This indicates a higher initial solvent content of the resin; ca. 22.5% which is greater than the resist and the resist analog formulations, which contained about 16% solvent. It also indicates that the resin tends to hold more tenaciously to the solvent than either the resist or the resist analog formulations. Thus both the resist and the resist analog can be prebaked at a lower temperature to achieve the same residual solvent content as for the resin film.

(2) For the resist and the resist analog formulations, a large percentage (ca. 80%) of the initial solvent content of the film is driven off between the prebake temperatures of 70°C and 100°C. In contrast, the resin film loses only about 50% of the initial solvent content in the same prebake temperature range.

For the P(MMA), the resin and the resist analog cases where almost all of the weight loss can be attributed to the solvent, the percent residual solvent content of the film after each prebaking can be calculated from the asymptote of Fig. 2 and the weight loss at each prebaking temperature. In the case of the resist formulations, the percent residual solvent content at each prebaking condition can be estimated by assuming further that at the asymptotic prebaking temperature of 160°C, the thermal decomposition of PAC is complete and thus all of the N_2 by-product of the thermal decomposition is evolved. This assumption should yield only minor errors in the calculation of the percent residual solvent for the resist formulation, because the maximum weight which can be attributed to N_2 evolution is only about 1.4% of the dried film. The percent residual solvent can thus be computed as follows:

$$\% \text{ Solvent} = 100\left(\frac{W_s - \Delta W}{W_T - \Delta W}\right) \tag{1}$$

where W_s, ΔW and W_T are the solvent content before prebaking, the weight loss after prebaking and the total film weight before prebaking.

Figure 3 shows the solvent content of the cast films at various prebaking temperatures. Again the same conclusion can be drawn from Fig. 3 as from Fig. 2 and that there is a marked decrease in the ability of the resin to hold on to the solvent with the addition of either PAC or its analog, the PACA. The question arises whether this phenomena can be explained purely on the basis of simple dilution of the resin by the PAC (the resist being 82% resin and 18%, PAC). For example, the PAC acts as an inert diluent of the resin in regard to the solvent retention characteristic of the resist. If this is indeed the case, the solvent content of the resist when adjusted for its lower resin content using a simple dilution calculation should agree with the experimental solvent content of the pure resin. The adjusted solvent content of the resist (16% ÷ 0.82) is only 19.5% compared to the experimental value of 22.5% for the pure resin. This discrepancy could be due to a strong interaction between the resin and the PAC which reduces the tendency of the resin to retain the solvent.

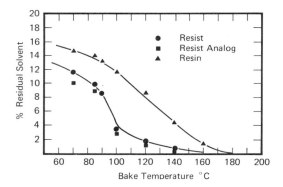

Fig. 3. Solvent content versus baking temperature of casted resin, resist and resist analog films on silicon substrate. Baking time was 30 min.

In Fig. 4 are the dissolution rate versus the prebake temperature plots for the three different films studied. Three conclusions are readily apparent from the plots:

(1) Between 70°C and 140°C, the resin has ca. 2 orders of magnitude.

(2) The solubility rate of the resist does not decrease monotonically with prebake temperature.

(3) The solubilities of all three films are approximately equal at the 160°C prebake temperature, despite their large differences at the lower prebake temperature.

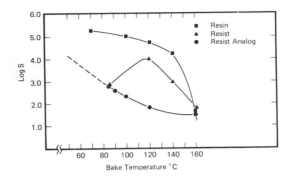

Fig. 4. Dissolution rate versus baking temperature plots of casted resin, resist and resist analog films on silicon substrate. Baking time was 30 min.

The increase in the solubility rate of the resist formulation between 85°C and 120°C prebake temperatures can be attributed mainly to the decomposition of the PAC yielding N_2 and indene acid. Gritter and Dawson,[5] using FT-IR techniques, have shown that the PAC starts to decompose rapidly at a prebaking temperature of ca. 100°C. The decomposition of the PAC not only increases the equilibrium solubility of the resist by forming the more soluble by-product, the indene acid, but the evolution of N_2 gas also facilitates solvent permeation into the glassy resist via formation of void, stress and microcracks in the resist film associated with N_2 evolution.

Increase in prebaking temperature from 120°C to 160°C results in a drastic decrease in the solubility rate of the resist formulation. This could be due to the fact that the N_2 evolution has ceased (all of the PAC has decomposed) at this higher prebake temperature, the voids created by the evolution of N_2 has been annealed, and the residual solvent has been reduced to a very low level. It could be argued that the sharp decrease in the solubility rate on prebaking beyond 120°C is the result of extensive crosslinking of the resist, forming exceedingly high molecular weight. Thus the molecular weights of the resin prebaked at 70°C, 140°C and 160°C were measured using a low angle laser light scattering photometer. The weight average molecular weights ($\overline{M}_w$) obtained for the resin prebaked at 70°C, 140°C and 160°C were 6000, 5000 and 7000, respectively. These results indicate very little change, if any, in the molecular weight during the prebaking.

The large difference in the dissolution rate-prebake temperature plots between the resin and the resist or the resist analog could perhaps be explained by the difference in the residual solvent content between the resin and the resist films prebaked at the same temperature. A solubility rate comparison between the resin and the resist or the resist analog film based on solvent content is shown in Fig. 5. The curves in Fig. 5 show that at lower solvent content (<2%), the solubility rate between the resin and the resist or resist analog are nearly equal. However, a large difference in solubility rate between the two films exists at higher residual solvent content. It is interesting that the large differences in the solubility rate between the resin and the resist analog develops in a relatively narrow solvent content range, i.e., between 1.5% to 4.5%. At the higher residual solvent content, the slope of the resin and the resist analog plots are essentially the same. This suggests that at high solvent content (>4%), the films are sufficiently swollen that the rate of developer penetration becomes too fast to be the controlling rate of the dissolution process. In this swollen state of the resist, it is the equilibrium properties, i.e., the free energy of mixing (or dissolution), that become the controlling factor. However, as the percent residual solvent decreases (<4%) and the degree of annealing of the films increases, the contribution of the developer transport properties in the resin or resist glass becomes more important, until it predominates when the residual casting solvent reaches ca. 1.5%.

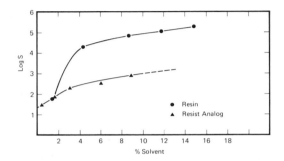

Fig. 5. Normalized solubility rate versus the volume fraction of solvent in the resin and resist analog.

The free volume concept in interpreting the diffusion coefficient of the solvent molecules in glassy polymers was developed by Fugita.[6] He showed that the diffusion coefficient of solvents in glassy polymer can be related to the volume fraction of the solvent in the following way,

$$\ell n \left[\frac{1}{\frac{D}{D_0}} \right] = K + \frac{m}{V_1} \tag{2}$$

where K and m are material constants related to the solvent-polymer pair D and D_0 are the diffusion coefficients in the swollen and the unswollen polymer, respectively. V_1 is the solvent volume fraction in the polymer.

From Eq. (2), it appears that for phenomena which depend on the free volume, e.g., the diffusion coefficient of solvents in a polymer, would have a relationship which would have an exponential relationship with V_1. Figure 6 shows a plot of the reciprocal of the natural logarithm of the normalized solubility rate (S/S_0) versus the reciprocal of the volume fraction of the solvent in the resist analog, and for the resin formulation. S and S_D are the solubility rate of the resin or resist analog in the swollen state and in the solvent free state, respectively. Although the data points are somewhat sparse, there is an indication that the curves for both the resin and the resist analog are linear, thus having a strong analogy to the dependence of the diffusion coefficient on V_1.

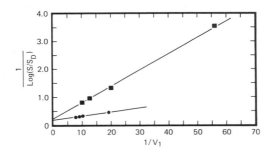

Fig. 6. Plots of the log reciprocal of the normalized solubility rate versus the reciprocal of the volume fraction of the solvent (V_1). $\square$—Resin, $\bullet$—resist analog.

It could be argued that the relationship between D and S described above is fortuitous. Previous papers,[7,8] however, show that S and D are uniquely related. In a computer simulation of the kinematics of polymer dissolution, it has been shown[8] that the solubility rate of glassy polymers does indeed depend strongly on the diffusion coefficient of the solvent in the polymer glass.

Figure 7 shows the drying curve of 1.5 μm thick P(MMA) on silicon substrate at ambient condition. The curve shows a relatively rapid initial solvent loss up to about 40 hours followed by very slow drying. If one were to plot these drying curves in semilog coordinates the resulting curve could be approximated by two exponentials of very different time constants. The fast segment of the drying curve show a time constant of 47 hours, while the slow segment is 10 times as long.

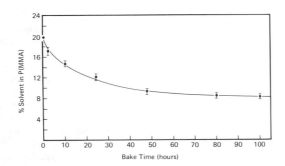

Fig. 7. Drying curve of 1.5 μm P(MMA) film on silicon substrate at ambient temperature $(T\sim25°C)$.

Figure 8 shows the drying curve of 1 μm thick P(MMA) on silicon substrate at various temperatures. Except for the 137°C+158°C, the shape of the drying curves is very similar to that of Fig. 7, e.g., a fast segment followed by very slow drying. The drying curve for the 85°C and 100°C baked samples are virtually similar, with fast and slow time constants of about 23 secs and 400 secs, respectively. The fast and slow time constants for the 117°C baking temperature drying curve are approximately 14 secs and 250 secs, respectively. The time constants for the 137°C and 158°C were too fast to be measured with reasonable precision.

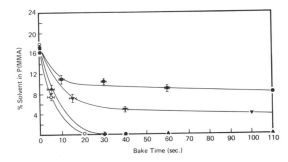

Fig. 8. Drying curve of 1.5 μm P(MMA) film on silicon substrate at various baking temperatures.
● – 85°C, □–100°C, ▼ – 117°C, ▲ – 137°C and ○ – 158°C.

The drying rate phenomena (the two exponential rates) observed here is perhaps related to the change in the free volume or the mobility of the P(MMA) chain segments as the sample traverses a certain solvent concentration range. At the higher solvent concentration, the P(MMA) chain segments may have enough mobility to facilitate solvent diffusion through the polymer matrix. At any given baking temperature, a concentration is reached whereby the solvent must diffuse through a glassy matrix, thereby reducing the solvent loss drastically. Another view is that of sorbed or strongly "bound" solvent on the P(MMA) molecule. At any temperature, there exists an equilibrium between sorbed and "freely diffusing" solvent molecule. Thus, as the concentration of solvent goes down, the concentration of "free" solvent molecule also goes down, thereby reducing the solvent evaporation rate.

Based on the dissolution curve of Greeneich,[1] and the drying curve of Fig. 8, it is apparent that the dissolution rate of P(MMA) cannot be simply interpreted in terms of residual solvent content. Figure 8 shows that at 150°C baking temperature all of the solvent is essentially driven out of the 1.5 μm P(MMA) film in about 20 secs, yet the dissolution rate after 5 min and 160°C bake is still very fast as is shown in Fig. 1. It takes at least 30 min baking at 160°C to show a marked reduction in the dissolution rate of P(MMA). The 30 min extra baking time at 160°C can perhaps be explained as the annealing time to allow the P(MMA) to "densify" after the thorough removal of the casting solvent. Because of the very rapid evaporation of the solvent at 160°C (C_6H_5Cl boils at 132°C), the dried P(MMA) have large "extra free volume" (microvoids, if you may). These "extra free volume", if not allowed to relax, will allow the solvent to diffuse very rapidly, resulting in very fast dissolution rate. This observation is consistent with the hypothesis offered[4] to explain the fast dissolution rate of P(MMA) which has been exposed to electron beam or X-ray radiation. Electron beam and X-ray radiation results in gaseous products which on evolution create "extra free volume" in the P(MMA) films.

IV. SUMMARY

The study of the effect of prebaking temperature on the residual solvent content and solubility rate of solvent-cast resin, resist, resist analog and P(MMA) have provided insights into the dissolution kinetics of two component resist materials and high molecular weight polymers. Some of the more important findings are:

- Solvent content in the prebaked resin has a gross effect on solubility rate, i.e., a few percent change in the solvent content can mean several orders of magnitude change in solubility rate.

- The addition of PAC or PACA to the resin results in lower solvent content in the cast film at any prebaking temperature, i.e., at 85°C prebake, pure resin contains ca. 14% solvent, while the resist or the resist analog contains only ca. 9.5% by weight.

- The dissolution kinetics of resin or resist films containing less than 4% residual solvent becomes a diffusion controlled process, while films containing greater than 4% are controlled by the equilibrium solubility of the resist or resin-developer pair.

- The strong dependence of the dissolution rate on residual solvent content can be interpreted in terms of a free volume concept in the same way that diffusion coefficient data have been treated.[7]

- At 158° baking temperature, the rate of solvent evaporation is very high, e.g., percent residual solvent in 1.5 μm P(MMA) film is reduced to about 0% in 20 secs of baking time, but the dissolution rate is also very high.

- The rapid evolution of the solvent at ~160°C leaves "extra free volume" and strain in the P(MMA). The dried film requires at least 30 min of annealing at 160°C to reduce the dissolution rate significantly. This observation is consistent with the hypothesis that the large increase in the dissolution rate of radiation exposed P(MMA) is the evolution of gaseous products, which creates extra free volume in the polymer matrix.

REFERENCES

Greeneich, J. S. (1975). J. Electrochem. Soc., 22, 970.
Dill, F. H., and J. M. Shaw (1977). IBM J. of Res. & Dev. 21, 2.
Ouano, A. C., and J. A. Carothers (1977). Unpublished data.
Ouano, A. C. (1978). Polymer Engineering and Sci. 18, 306.
Gritter, R. J., and B. L. Dawson, private communication.
Fugita, H. (1961). Fortschr. Hochpolym.-Forsch; Bd. 3, 1.
Ueberreiter, K. (1968). "The Solution Process," Diffusion in Polymers, J. Crank and G. Park, Editors (Academic Press, New York).
Tu, Y. O., and A. C. Ouano (1977). IBM J. of Res. & Dev. 21, 131.

USE OF EXCIMER FLUORESCENCE TO MEASURE POLYMER-SOLVENT INTERACTIONS[1]

Curtis W. Frank
Department of Chemical Engineering
Stanford University
Stanford, California 94305

ABSTRACT

Excimer fluorescence is an ubiquitous photophysical phenomenon in the aromatic vinyl polymers. A nominal excimer forming site (EFS) may be generated whenever two aromatic rings lie in a coplanar sandwich arrangement. Detection of the characteristic fluorescence from the excited complex allows an estimation of the number of ring-ring contacts. In the aromatic vinyl polymers such contacts may occur intramolecularly between rings on adjacent or nonadjacent repeat units or intermolecularly between rings on different polymer coils. In this study, the effect of solvent on intramolecular (adjacent) excimer formation is examined for the polymers poly(1-vinylnaphthalene) and poly(2-vinylnaphthalene) and the model compound, 1,3-bis(2-naphthyl)propane. An homologous series of alkyl benzene solvents is used in which there is a regular increase in molar volume as well as variations in solvent structure and flexibility. The stability of the excimer and the extent of collisional interaction with the surrounding solvent shell are shown to depend critically on solvent structure, the stability decreasing both as the solvent molar volume increases and as the solvent flexibility is reduced.

INTRODUCTION

Intramolecular excimer formation is a phenomenon interesting in its own right as a widely occurring photophysical process in low molecular weight aromatic compounds and aromatic vinyl polymers. A more important feature, however, may be the use of excimer fluorescence as a probe of molecular structure and mobility. Electronic stabilization of the excimer depends critically upon the overlap between the two aromatic rings comprising the excited complex. Since the complex will be solvated in solution, examination of the energetics of excimer formation should provide information on the solvent shell interaction.

---

[1]Editors' Note: This paper was previously published in Organic Coatings and Plastics Chemistry Preprints, 45, 433-438 (1981).

The polymers examined in this study are poly(1-vinylnaphthalene) (P1VN)
and poly(2-vinylnaphthalene) (P2VN). These were selected to illustrate
the influence of local steric interference between the aromatic rings
and the backbone on the spectroscopic behavior. In addition, 1,3-bis
(2-naphthyl)propane($\beta\beta$DNP) is examined as a model compound for excimer
formation in P2VN. It allows intramolecular interaction between
adjacent aromatic rings without structural interference from backbone
segments found in the polymer. The solvents are part of a homologous
series of alkyl benzenes including benzene (B), toluene (T), ethyl-
benzene (EB), m-xylene(X), n-propylbenzene (PB), iso-propyl benzene (C),
1,3,5-mesitylene (M), n-butyl benzene (BB) and n-pentyl benzene (AB).
Here the letters in parentheses serve to identify the solvents in the
figures. These solvents were selected because the solvent molar volume
increases regularly in the normal alkyl derivatives, albeit with a
change in the aromatic/aliphatic character. In addition, it is of
interest to compare the behavior of the normal alkyl derivatives with
the isomeric species of essentially the same molar volume.

EXPERIMENTAL

$\beta\beta$DNP was synthesized by Dr. R. Trujillo of Sandia Laboratories using
a slight modification of the procedure proposed by Chandross (1). It was
purified by multiple passes through a gel permeation column. The free
radical polymerization of P2VN was described previously (2) and the P1VN
sample was obtained from Polysciences. Both polymers were purified
by multiple precipitation. The alkyl benzene solvents were of highest
available quality and were distilled on a Todd Scientific Co. Precise
Fractionation Assembly Model A before use. All solutions were 5×10^{-5}
M and were bubbled with dry nitrogen gas until a constant excimer
intensity was achieved before measurement of the fluorescence spectrum.
The spectrometer has been described previously (2). Relative fluorescence yields for all room temperature spectra were analyzed by taking
integrated band areas.

RESULTS AND DISCUSSION

In general, three spectroscopic parameters may be used to characterize
the photostationary state fluorescence results. The first is the ratio
of excimer to monomer integrated band intensities, which is equivalent
to the ratio of relative fluorescence yields. Although this provides
important information on the overall efficiency of excimer formation,
the emphasis in this preprint is on the spectroscopic parameters of
emission band position and width of particular interest is the excimer
band position. ν_D. An approximate measure of the stabilization energy
of the excimer complex is the position of the excimer band relative to
the monomer ($\nu_M-\nu_D$). Since the monomer band appears to be independent
of solvent for all compounds examined, an increase in ν_D corresponds
to a decrease in stability. A more accurate treatment of the data including binding energy measurements made at high temperatures leads to
the same trends. The third parameter is the excimer band width which is
a measure of local solvent interactions with the complex. It is well
known that dispersive interaction and collisions tend to broaden
absorption and emission bonds in solution relative to the vapor phase
spectra. A similar effect is expected for the excimer complex.

The spectroscopic results for the ββDNP model compound will be considered first. In order to isolate the influence of solvent structure on the excimer band position, νD must be corrected for changes in solvent density and refractive index. To do this we use the expression $\Delta\nu = A\Delta f$ where $\Delta\nu$ is the change in excimer band position relative to benzene solution in units of

cm^{-1}, A is a constant and Δf is the change in the quantity $f = \rho \left[\dfrac{n^2-1}{2n^2+1} \right]$

in which ρ is in units of g/cm^3. Since it appears that the solvent shape and size may have a significant influence on the excimer formation and stability, the corrected νD values are plotted against solvent molar volume, V_m, in Figure 1. To be sure, V_m is a rather crude measure of solvent size. More informative parameters may be Pitzer's acentric factor or the quantity V_o/V_f in which V_o is the occupied volume in the pure solvent and V_f is the free volume in the pure solvent. Use of these parameters does not change the qualitative trends, however.

An important observation from Figure 1 is that there are no significant variations in stability within the groups of solvents with similar molar

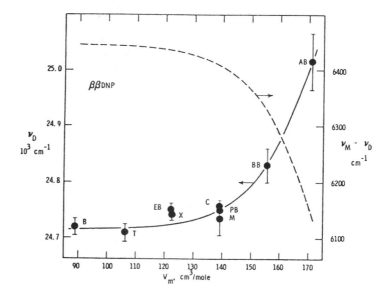

Figure 1: Dependence of excimer band maximum and separation between monomer 0-0 band and excimer band maximum on solvent molar volume for 1,3-bis (2-naphthyl)propane. The solvent symbols are defined in the text.

volume: (ethyl benzene and m-xylene) and (n-propyl benzene, iso-propyl benzene and 1,3,5 mesitylene). The significant feature of the results is the increase in band position, or decrease in complex stability as indicated by the dashed line for $\nu_M - \nu_D$, beginning with the propyl benzene isomers and rapidly increasing thereafter for the long chain butyl and pentyl derivatives.

The second set of data from which to elucidate the local solute-solvent interactions is that for the half width of the excimer band. These results are plotted against V_m in Figure 2. As was observed for ν_D, there is little change in $\Delta\nu_{1/2}$ until the propyl benzene isomers are reached. At this point a broadening is observed which rapidly increases with further increases in solvent size. This broadening must indicate that there is increased interaction of the excimer complex with the surrounding solvent cage. It is interesting to note that the half widths for 1,3-bis (1-naphthyl)propane ($\alpha\alpha$DNP) and $\beta\beta$DNP are essentially identical. In ethanol $\Delta\nu_{1/2}$ for $\beta\beta$DNP was estimated to be 4380 ± 50 cm^{-1} which agrees well with $\beta\beta$DNP in the smaller alkyl benzene solvents. Thus, it appears that solvent cage collisional interactions with the ring complex are not very dependent upon the relationship of the rings to the backbone for the model compounds. As will be shown, this situation is considerably different for the polymers.

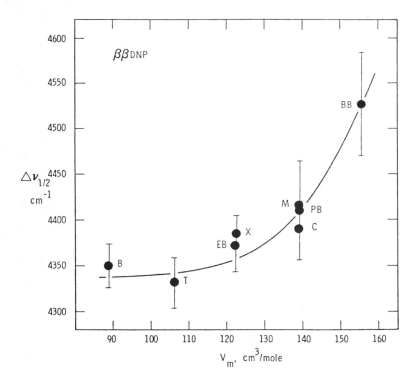

Figure 2: Dependence of excimer band width on solvent molar volume for $\beta\beta$DNP.

For example, the corrected values of ν_D are plotted in Figure 3 for both polymers. The solid lines are simply smooth curves through the normal alkyl derivatives. The rectangular dashed line boxes enclose data for solvent isomers of essentially identical molar volume. In both P1VN and P2VN there is an increase in ν_D with increasing molar volume with the increase most pronounced after the propyl benzene isomers. This behavior is identical to that of $\beta\beta$DNP where the effect is attributed to reduction of available free volume and resulting increase in steric hindrance. Whereas ν_D for the isomers is essentially equal for $\beta\beta$DNP, however, significant differences arise for the polymers. Replacement of single long chain substituents with multiple short chain groups, while maintaining constant molar volume, has the effect of increasing ν_D.

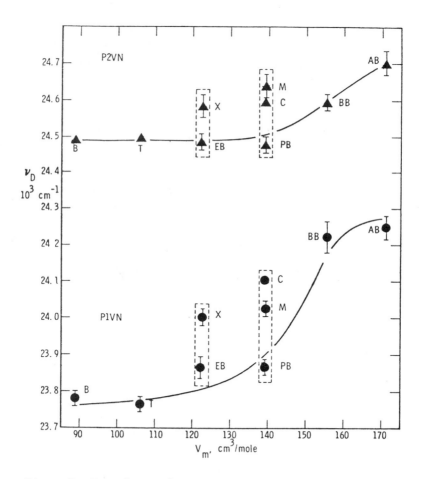

Figure 3:   Dependence of excimer band maximum on solvent molar volume for P1VN and P2VN.

The trend toward decreasing complex stability with increasing solvent
size or decreasing group flexibility is apparent in all solvents.   The
polymer is somewhat less stabilized than the model.   This difference is
110 cm^{-1} ± 25 cm^{-1} for all normal derivatives and is considerably larger
at 220 ± 40 cm^{-1} for the multiply substituted isomers.   No results for
$\alpha\alpha$DNP in the alkyl benzene series are available for comparison with P1VN.
Data from Avouris (3), however, indicate that (ν_M, ν_D) values for $\alpha\alpha$DNP
in ethanol, when corrected for density and refractive index differences, lead
to estimated values of (29400, 23400) in benzene, yielding $\nu_M - \nu_D$ = 6000 cm^{-1}.
As was the case for 2-substitution, this is more stabilized than P1VN since
(ν_M, ν_D) values for P1VN in benzene are (29400, 23780), yielding $\nu_M - \nu_D$ =
5620 cm^{-1}.

The band width results for both polymers are shown in Figure 4.   Here
the solid lines are smooth curves through the normal solvents, the rectangles

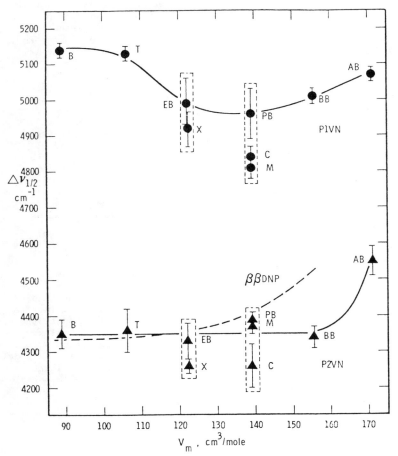

Figure 4:   Dependence of excimer band width on solvent molar volume
            for P1VN and P2VN.

enclose solvent isomers of similar molar volume and the dashed line
represents the smoothed data for the model compound ββDNP. Trends
within each solvent series for the polymers will be considered first
followed by comparisons with the corresponding model compounds. Although
basically similar individual solvent effects are obtained, there are
some differences for the two polymers. The normal derivatives for P2VN
show essentially no change until the bulky n-pentyl benzene is reached, at
which point there is an increase. On the other hand, the P1VN solutions
show decreasing band widths up to n-propyl benzene, after which there is
a broadening similar to that found for P2VN. In addition, the multiply
substituted solvents all show decreases in band width with respect to
their normal counterparts, but the effect is somewhat larger for P1VN
than for P2VN. From comparison with model compound results we note also
that half widths for ββDNP are similar to those for P2VN. In fact, all
of the data, with the exception of m-xylene, isopropyl benzene and butyl
benzene, are within the observed error limits. In those solvents in which
deviations do occur they are generally narrower for the polymer than
for the model compound. The situation is quite different for 1-substi-
tution. From Avouris' results the band width for ααDNP was estimated to
be approximately 4000 cm^{-1}; it is clear from Figure 4, however, that $\Delta\nu_{1/2}$
for the polymer is much larger.

The second point of major concern is the large difference between the
widths for P1VN and P2VN. To explain this, we note that excimer band
broadening may result from both solvent-complex and from backbone-complex
interactions. The observation of the same half widths for ββDNP and P2VN
in the small molar volume solvents indicates that there must be a similar
solvent cage structure in both the model and the polymer. This may be
expected because the ring complex is directed away from the chain backbone,
allowing for easier solvation and minimum steric interference. Backbone
interference in the model compound ααDNP is certainly present to some extent
in view of the stabilization energies. The influence on the band width is
small, however, with $\Delta\nu_{1/2}$ being only slightly larger for ααDNP than for
ββDNP. The explanation for the large polymer value must involve extensive
backbone interference with the excited ring complex.

To summarize, the excimer complex serves as a unique probe for the
measurement of structural interaction in solution. If attention is
focussed on the complex surrounded by a solvent shell of unspecified
geometry, one expects optimum packing and minimization of repulsive forces
to be reached in the case of the model compound. If, however, the complex
is situated in the middle of a polymer chain, solvent packing must inevitably
be somewhat different and most probably more restricted to accomodate the
adjacent chain segments. The lower stability of the polymeric complex which
is observed is then a reflection of this more crowded configuration.
Furthermore, the larger destabilization of the polymer relative to the model
compound for 1-substitution as compared to 2-substitution may be related
to the steric hindrance expected between the ring hydrogen in the 8 position
and the chain backbone. No such interference exists for 2-substitution
as the ring is directed away from the backbone.

REFERENCES

1. E. A. Chandross and C. J. Dempster, J. Am. Chem. Soc. <u>92</u>, 3586 (1970).

2. C. W. Frank and L. A. Harrah, J. Chem. Phys. <u>61</u>, 1526 (1974).

3. P. Avouris, J. Kordas and M. A. El Bayoumi, Chem. Phys. Lett. <u>26</u>, 373 (1974).

ACKNOWLEDGEMENTS

This work was supported by the US Energy Research and Development Administration while the author was employed by Sandia Laboratories, Albuquerque, New Mexico and by the Stanford Institute for Energy Studies.

PROCEDURE FOR PREDICTING THERMODYNAMIC PROPERTIES OF POLYMER SOLUTIONS FROM DATA ON THE SOLVENT AND OLIGOMERS

Maurice L. Huggins

135 Northridge Lane, Woodside, CA   94062

ABSTRACT

A general procedure is described, for calculating the thermodynamic properties of polymer solutions from accurately measurable properties of the pure liquid solvents, pure liquid oligomers, and solutions of the oligomers in the solvents. This procedure is based on the author's quantitatively accurate theories for the energies and densities of pure simple liquids and their mixtures.

KEYWORDS

Thermodynamics of liquids; thermodynamics of solutions; polymer thermodynamics; oligomers.

INTRODUCTION

The solubilities and other thermodynamic properties of polymer solutions are exact functions of the dependence of the enthalpy and entropy on the concentration, temperature and pressure. The energy, enthalpy and volume of a solution depend largely on the interactions between close-neighbor solvent molecules and polymer mers, hence on the chemical compositions of these molecules and mers. To understand and to predict accurately the thermodynamic properties of polymer solutions we must therefore understand and be able to predict the dependence of these properties on the chemical compositions and sizes of the molecules and mers.

Although some progress on this problem, with respect to a few polymer-solvent systems, has been made, no procedure of general applicability has (to my knowledge) been previously proposed.

Both the enthalpy and the entropy are important, but in this presentation I shall deal primarily with the enthalpy aspects.

To simplify the problem somewhat, I assume that the external pressure is not larger than about one atmosphere. The enthalpy and energy changes, with which we are concerned, are then practically equivalent. Also, I neglect possible changes in the *intra*molecular energies and enthalpies, on forming a solution from its components. Moreover, I neglect possible small changes in the energy or enthalpy of interaction between atoms or groups that are not in mutual contact. I treat the intermolecular energies of the liquids as the sums of the "contact energies" between close units (molecules or mers).

PURE LIQUIDS

I have recently developed and tested a theory (Huggins, 1980a, 1980b) of the thermodynamic properties of pure liquids, dealing first only with *simple* liquids, for which one can reasonably assume the energy-volume-temperature-pressure properties to be interrelated as they would be for a hypothetical model liquid in which (at a given temperature) each molecule makes the same number of contacts with others and the contacts all have the same "contact energy" and the same "contact distance". Contact energies, like bond energies, are taken as positive, whereas the molal energies (E) are negative.

The contact energy for Ávogadro's number of contacts, between molecules of type α, is assumed to be given by the equation

$$D_{\alpha\alpha} = \left[\frac{c_{\alpha\alpha}}{(r_{\alpha\alpha} - b_{\alpha\alpha})^6}\right] - D^* \exp\left[a(r^*_{\alpha\alpha} - r_{\alpha\alpha})\right] \tag{1}$$

Following many previous researchers working on this and similar problems, the attraction energy, represented by the first term on the right side of this equation, is assumed to be of the London (1937) form, with proportionality to the inverse sixth power of the contact distance. I subtract from this distance, however, a constant, $b_{\alpha\alpha}$, that is characteristic of the liquid. This allows for the fact that the oscillating charges (responsible for the attraction) are not located at the molecular centers.

The other term in this equation represents the attraction energy, in the exponential form successfully used for ionic crystals (Born and Mayer, 1932; Huggins and Mayer, 1933; Huggins, 1937). D^* and "a" are general constants, assumed to be the same for all liquids. The other parameters in the equation ($c_{\alpha\alpha}$ and $r^*_{\alpha\alpha}$) are characteristic of the liquid and independent of the temperature and pressure.

The molal contact energy ($D_{\alpha\alpha}$) is related to the molal intermolecular energy (E_α) and the (temperature-dependent) contact number (σ^0_α) by the equation

$$D_{\alpha\alpha} = \frac{-2E_\alpha}{\sigma^0_\alpha} \tag{2}$$

E_α is the negative of the energy of vaporization of the liquid to an infinitely dilute gas:

$$E_\alpha = -\Delta E^0_{vap,\alpha} = RT - \Delta H^0_{vap,\alpha} \tag{3}$$

The intermolecular contact distance, $r_{\alpha\alpha}$, is assumed to be related to the molal volume (V_α), the density (ρ_α), the molecular weight (M_α), and the contact number (σ^0_α) by the equation

$$r_{\alpha\alpha} = f_\sigma V_\alpha^{1/3} = f_\sigma\left(\frac{M_\alpha}{\rho_\alpha}\right)^{1/3} \tag{4}$$

where

$$f_\sigma = 1.32917 - 0.00035 \ (12 - \sigma^0)^{3.36} \tag{5}$$

is the equation for a smooth curve agreeing with the known geometrical relation for crystals in which each atom (or molecule) is surrounded equidistantly by 6, 8 or 12 others. See Fig. 1.

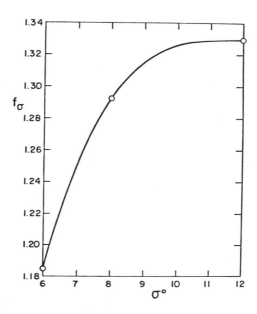

Fig. 1. Dependence of f_σ on σ^0. From Huggins (1980a).

Since the maximum contact number of spheres is 12, the curve is assumed to approach that number asymptotically.

The molal volume at a given temperature is a function of the compressibility. The limiting compressibility, as the pressure approaches zero, is given by the relation

$$\beta_{\alpha,P=0} = \frac{1}{V_\alpha} \left(\frac{\partial V_\alpha}{\partial P} \right)_T \tag{6}$$

Assuming

$$D^* = 10^5 \ \text{Joules} \tag{7}$$

and using apparently accurate experimental values (over considerable temperature ranges) of the vaporization energy, density and compressibility, the parameters listed at the top of Table I have been deduced.

TABLE I: Parameters and Properties for Liquid Benzene and Cyclohexane

	Benzene	Cyclohexane
	Temperature-independent parameters	
$D* / 10^5 J$	(1)	(1)
$a / 10^8 cm^{-1}$	2.400	2.400
$b / 10^{-8} cm$	1.984	2.0
$c / 10^{-42} J cm^6$	40.97	58.1773
$r* / 10^{-8} cm$	4.600	4.8
	25°C values for temperature-dependent functions	
$E / 10^3 J$	-31.36051	-30.5611
$D / 10^3 J$	7.375481	7.337599
V / cm^3	89.4020	108.749
σ^0	8.503991	9.525
$r / 10^{-8} cm$	5.838392	6.212504
$D_{attr} / 10^3 J$	12.49484	
$D_{rep} / 10^3 J$	-5.11936	
$G / 10^3 J$	-83.066	
$\rho / g\, cm^3$	.873732	.77390
$\Delta H^0_{vap} / 10^3 J$	33.83943	33.040
$\beta / 10^{-5} cm^3 J^{-1}$	97.000	113.979

Using these parameters, the magnitudes of the parameters and properties given (for 25°C) in the lower part of the table have been calculated. Other values over the whole liquid range at 5° intervals have also been computed. These magnitudes agree *quantitatively*, with the experimental measurements, over the ranges of those measurements. See Fig. 2 for graphs showing the temperature dependence of the contact parameters for liquid benzene.

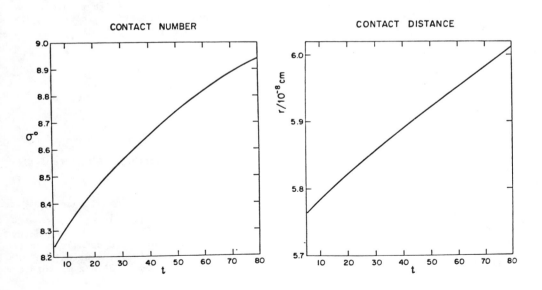

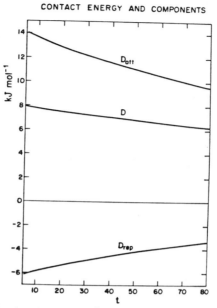

Fig. 2   Contact parameters for benzene.   From Huggins (1980a)

Application of the theory to carbon tetrachloride is now in progress and many other simple liquids can doubtless be dealt with similarly, if (or when) accurate vapor- ization energy and density data are available.

Oligomers having mers that are sufficiently uniform chemically (with regard to their intermolecular contacts) would be good subjects for treatment in this way. I suggest studies of the following, each with several values of n:

$$H(CH_2)_n H$$

$$H\left(\begin{array}{c} CH-CH_2 \\ | \\ CH_3 \end{array}\right)_n H$$

$$H\left(\begin{array}{c} CH-CH \\ | \\ C_6H_5 \end{array}\right)_n H$$

Extrapolation of the oligomer parameters should lead to parameters characteristic of the mers in higher polymers.

MIXTURES OF SIMPLE LIQUIDS: EXCESS ENTHALPIES

The literature contains many reports, especially during the last 15 years, of mea- surements of the concentration dependence of the enthalpy, volume and Gibbs energy as two low-molecular-weight substances are mixed. The results are usually reported as excess enthalpies, excess volumes, and excess Gibbs energies, measuring the changes in these properties when the two substances are mixed to give a mixture con- taining one mole of mixture.

I have been concerned primarily with the excess enthalpy data, developing and testing a theory closely related to the theory just described for pure simple liquids.
(Huggins, 1970, 1975, 1976a, 1976b, 1977)

Three parameters are involved:

(1) an energy parameter, ε_Δ, measuring the energy change when two contacts between
like molecules are replaced by two contacts between unlike molecules;
(2) a contact ratio parameter, $r_\sigma = \sigma_2^0 / \sigma_1^0$, equal to the number of contacts made
by one kind of molecule, divided by the number of contacts made by the other kind
of molecule; and
(3) an equilibrium constant K, measuring the relative probabilities of the three
kinds of contacts.

The magnitudes of the three parameters can be deduced from the shapes of the experimental curves - if the experimental data are accurate.

Testing the theory by applying it to published data showing very small standard
deviations (of the experimental data points from a smoothing curve of the ideal
form), deducing the theoretical parameters, and then comparing the theoretical
curve (calculated from those parameters) with the smoothing curve, shows *quantitative* agreement. See Fig. 3.

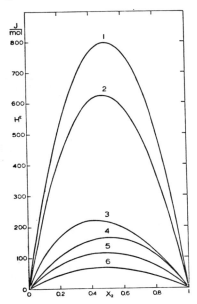

Fig. 3.   Theoretical and experimental excess enthalpy curves (indistinguishable, on
          this scale):   (1) benzene + cyclohexane, (2) cyclohexane + toluene, (3)
          cyclohexane + n-hexane, (4) cyclohexane + CCl₄, (5) benzene + CCl₄, (6)
          benzene + toluene.   From Huggins (1975).

I have attempted to calculate excess enthalpy parameters from about 50 reports of
experimental data for solutions of normal alkanes in benzene, cyclohexane and carbon tetrachloride, but very few of these sets of data are accurate enough to serve
as a basis for extrapolating to higher polymers.   The standard deviations of the

experimental data are too large, different reports on the same system do not agree, and calculations of the same parameter from data for different alkanes in the same solvent do not fall on smooth curves, when plotted against the number of carbon atoms in the alkane molecule. The best data for my purpose appear to be for alkane solutions in benzene, but they are not good enough. See Fig. 4.

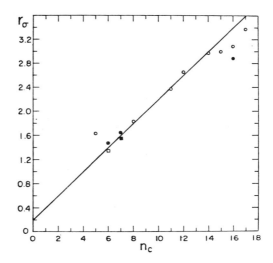

Fig. 4.   A plot of the contact number ratio (calculated from inadequate data)
          against the number of carbon atoms in the alkane molecule, for normal
          alkanes in benzene. From Huggins (1977).

The measurements should be made on mixtures of the pure components, with accuracies such as have been achieved in various laboratories for many other systems. There should be researches on many other oligomer + solvent systems and at temperatures other than the usual 25°C. Values of the important χ_h parameter, corresponding to the excess enthalpy values, should be calculated and plotted against concentration in weight fractions, as is often done for polymer solutions. *Such researches constitute an important part of the program I advocate.*

INTRAMOLECULAR CONTACTS

Flexible long chain polymer molecules, both in the pure polymers and in solutions with other molecules, make many intramolecular contacts. Experimental excess enthalpies depend, in part, on the relative numbers of intramolecular contacts per polymer molecule in the pure polymer and in the solution. The relative number in the solution is obviously a function of the average number of mers in the polymer chains, the concentration of the solution, the temperature, and the relative probabilities of the three types of contact (measured by the equilibrium constant K, already mentioned as one of the parameters determining the shape of the excess enthalpy curves).

An attempt has been made (Huggins and Kennedy, 1979) to use random walk statistics as an aid in extrapolating oligomer-solvent enthalpy properties and predicting properties of solutions of longer polymers in the same solvent. Because of the unreliability of the oligomer-solvent data and some questionable assumptions, our conclusions or similar conclusions based on the same assumptions should not be accepted without further checking. Also, the procedure we used would not be expected to be applicable to solutions for which the constant K is not equal to unity. *An important part of the program I advocate is the development of a better theoretical treatment of the effect of intramolecular contacting on the properties of polymer-solvent systems. With or without such a theory, there should be appropriate experiments, especially using well characterized polymers in the intermediate range of chain lengths. These researches should lead to procedures for predicting the enthalpic behavior of high polymer solutions from experiments on solutions of oligomers in the same solvent.*

ENTROPY EFFECTS

Although I have here neglected consideration of departures of the actual entropy changes on forming polymer solutions from the changes predicted (assuming random mixing) by me (Huggins, 1942) and by Flory (1942), I realize that these departures often contribute importantly to the Gibbs energy changes on mixing and so to important polymer problems - such as those involving equilibria between phases.

I have discussed the origins of these departures and possible procedures for evaluating their magnitude elsewhere (Huggins, 1971, 1972) and now merely want to suggest that *these and other theoretical ideas relating to entropies of mixing might well be tested by appropriate experiments on oligomer solutions.*

CONCLUSION

I have outlined a program that can (I believe) lead to a much better understanding of the properties of polymer solutions. Although I am not now in a position to contribute much to this program, I hope that others who can do so will see its importance and act accordingly.

REFERENCES

Born, M., and J. E. Mayer (1932). Z. Physik, *75*, 1–18.
Flory, P. J. (1942). J. Chem. Phys., *10*, 51–61.
Huggins, M. L. (1937). J. Chem. Phys., *5*, 143–148.
Huggins, M. L. (1942). J. Phys. Chem., *46*, 151–158.
Huggins, M. L. (1970). J. Phys. Chem., *74*, 371–378.
Huggins, M. L. (1971). J. Phys. Chem., *75*, 1255–1259.
Huggins, M. L. (1972). J. Paint Tech., *44*, 55–65.
Huggins, M. L. (1975). In C. E. H. Bawn (Ed.), *International Review of Science, Physical Chemistry, Series Two, Vol. 8*, Butterworths, London, Chap. 3, pp. 123–152.
Huggins, M. L. (1976a). J. Phys. Chem., *80*, 1317–1321.
Huggins, M. L. (1976b). J. Phys. Chem., *80*, 2732–2734.
Huggins, M. L. (1977). Br. Polym. J., *9*, 189–194.
Huggins, M. L. (1980a). J. Macromol. Sci.-Phys., *B18*, 409–421.
Huggins, M. L. (1980b). Colloid and Polym. Sci., *258*, 477–482.
Huggins, M. L. and J. W. Kennedy (1979). Polym. J., *11*, 315–322.
Huggins, M. L. and J. E. Mayer (1933). J. Chem. Phys., *1*, 643–646.
London, F. (1937). Trans. Faraday Soc., *33*, 8–26.